Stereochemistry of
Coordination Compounds

Inorganic Chemistry
A Textbook Series

Alexander von Zelewsky
Curriculum Vitae

Alexander von Zelewsky was born in 1936 in Zürich, Switzerland, and studied chemistry at the Swiss Federal Institute (ETH), Zürich. He graduated in 1960 (Dipl. Ing. Chem.) and obtained his PhD with Walter Schneider in 1964. He was a Miller Research Fellow at the University of California, Berkeley, from 1965 to 1967, working with Robert E. Connick. After returning to the ETH for 2 years, he became a full professor at the University of Fribourg, Switzerland, in 1969. During 1994 he spent a sabbatical at UC, Berkeley. His main research interests are the synthesis of coordination compounds for photochemical purposes and EPR spectroscopy of coordination compounds.

Stereochemistry of Coordination Compounds

Alexander von Zelewsky

University of Fribourg, Switzerland

JOHN WILEY & SONS

Chichester · New York · Brisbane · Toronto · Singapore

Other Wiley Editorial Offices

John Wiley & Sons, Inc., 605 Third Avenue,
New York, NY 10158-0012, USA

Jacaranda Wiley Ltd, 33 Park Road, Milton,
Queensland 4064, Australia

John Wiley & Sons (Canada) Ltd, 22 Worcester Road,
Rexdale, Ontario M9W 1L1, Canada

John Wiley & Sons (SEA) Pte Ltd, 2 Clementi Loop #02-01,
Jin Xing Distripark, Singapore 0512

Library of Congress Cataloging-in-Publication Data

Zelewsky, Alexander von.
    Stereochemistry of coordination compounds / Alexander von
Zelewsky
        p.   cm.—(Inorganic chemistry)
    Includes bibliographical references and index.
    ISBN 0-471-95057-2 (cloth   :   alk. paper).—ISBN 0-471-95599-X
(paper   :   alk. paper)
    1. Coordination compounds. 2. Stereochemistry. I. Title.
II. Series: Inorganic chemistry (John Wiley & Sons)
QD474.Z45   1995
541.2′23—dc20   95–19754
                                                           CIP

British Library Cataloguing in Publication Data

A catalogue record for this book is available from the British Library

ISBN 0 471 95057 2 (cloth)
ISBN 0 471 95599 X (paper)

Typeset in 10/12pt Times by Alden Multimedia, Northampton

Contents

To

Hedi

Preface

The Stereochemistry of Coordination Compounds is as old as Coordination Chemistry itself. Werner derived his theory of coordination, published in 1893, to a large extent from stereochemical arguments. The stereochemistry of species with various coordination numbers and geometries has since developed considerably. However, it has not yet reached the degree of elaboration of organic stereochemistry. Recent progress in several fields of coordination chemistry shows that we are in a period where stereochemical considerations are again increasing in importance. Model compounds for bioinorganic species, catalysts for enantioselective reactions, the building blocks for supramolecular assemblies and coordination units with well defined functionalities (molecular devices) all require an understanding and control of the stereochemistry of metallic centers.

I became greatly interested in the stereochemistry of coordination compounds when we were confronted with the problem of synthesizing well defined structures for photochemical molecular devices. A literature search for textbooks on the topic yielded a series of titles that partially covered the subject. These texts are, however, mainly concerned with issues that I call '*metric stereochemistry*' in this book. Inorganic chemistry textbooks generally also treat the other aspects that I call '*topographical stereochemistry*,' but in most cases only the basic concepts are given. We therefore came to the conclusion that a specialized treatment of this subject would be useful for both students and researchers in coordination chemistry alike.

Metric stereochemistry is treated only in a cursory way in this book, since there is no need to elaborate further in view of the excellent textbooks already available. The main emphasis is on the 'topographical stereochemistry' of several coordination numbers and geometries. Consequently OC-6, the octahedral coordination, being the most frequent geometry for metallic centers, takes the largest part of these considerations. The book is organized systematically along the lines of stereochemical properties and not by any form of systematic chemical considerations. It is therefore *not* intended to be a coordination chemistry textbook, based on stereochemical ideas. Chapters 1 to 6 treat stereochemistry mainly from a static point of view, while chapter 7 gives an outlook on topographical stereochemical descriptions of several kinds of reactions and molecular rearrangements. However, the whole field of the stereochemistry of transformations is one of great complexity and could easily be the subject of a separate book.

The scope of the book is largely confined to classical coordination complexes. I have made no attempt to encompass the enormously large field of organometallic chemistry (including cluster compounds), since it is my feeling that a systematic treatment of this subject from a stereochemical point of view would require a much greater involvement with the questions of connectivities, i.e. topological questions. Topographical stereochemistry can be treated from a highly abstract and formalistic point of view that requires corresponding mathematical methods. This book approaches the subject on a descriptive basis that makes use of our ability to deduce many stereochemical facts just by looking in the literal sense at a molecular model. The viewing of molecular models is a skill that can be trained to a high degree and it is the aim of the book to help the reader to develop an intuitive and also systematic way of doing this.

In order to give the reader the possibility of making more detailed studies, the book is fully referenced. I do not pretend in any way, however, that the references represent all or even the most important contributors to the stereochemistry of coordination compounds made so far. The selection, necessary for a textbook, is always to a certain degree arbitrary, so I apologize to all colleagues who might miss their own, or other, contributions that they think would have been worthy of inclusion. Those who have written a book themselves will forgive me, because they will certainly appreciate the difficulty. I am very happy that the publisher agreed to make the illustrations of the book available on the internet, where a home page can be reached under <http://sgich1.unifr.ch/ac/avzbook.html/>.

Although this is a single-author book, it could not have been written without the help of many people. I especially thank Professor Ken Raymond of UC Berkeley, who was an excellent host during my sabbatical in California in 1994 where most of this book took shape. I thank also Professor Richard Keene of James Cook University (Australia), who read the manuscript carefully and made useful comments and corrections. Dr Liz Kohl of Fribourg University managed all the computer files and references for the book from the beginning. Even when the files, traveling by Internet from Berkeley to Fribourg, were messy, they came back nicely organized the same day so that I could continue with my work. My students Véronique Monney and especially Marco Ziegler helped greatly with the figures and I thank them for their careful and competent contributions. In addition, Marco read the manuscript very carefully and I am greatly indebted to him for pointing out several inconsistencies and errors and for highly constructive suggestions for improvements. Although the finished book is the work of many, it is of course my own responsibility if there are still errors, omissions and other imperfections in the text. Since I am sure that there are points to be criticized and commented on, I would appreciate it if colleagues around the world would communicate their views to me. This is done nowadays with the least possible formalities by electronic mail. Please send comments to <Alexander.Vonzelewsky@unifr.ch>.

Fribourg, Spring 1995   Alexander von Zelewsky

1 Introduction

1.1 The Emergence of Stereochemistry of Coordination Compounds and its Present Role in Chemistry

The coordination theory formulated by Alfred Werner in an epoch-making publication in 1893 [1] is largely based on stereochemical arguments. It was the first generalization of stereochemical concepts in chemistry after the introduction of structural ideas in organic chemistry by Le Bel and van't Hoff 19 years earlier. Werner uses as one of the very basic facts, from which he deduces the octahedral coordination geometry of many metals, the number of isomers in several chromium(III), cobalt(III), platinum(IV), and platinum(II) complexes. Figure 1.1 is a reproduction of the first representation of an octahedron in the chemical literature, taken from that publication. The formula for the so-called compounds of higher order, by S. M. Jørgensen, which were based on the unitarian theory of valence (to each element was attributed a fixed valency number, which was thought to represent the number of bonds, the atom can form with other atoms of the same or another element) contained no real stereochemical information whatsoever. Werner realized clearly that an octahedral surrounding of a central atom would explain the two isomers of complexes $[M(NH_3)_4X_2]$ and also of $[M(NH_3)_3X_3]$ (Figure 1.1). One should realize that all this happened before the discovery of any internal structure of the atom.

The importance attributed to stereochemistry by Werner is shown by the fact that his first book, published in 1904, is devoted entirely to stereochemistry [2], although it deals mainly with non-metallic elements. In his next book, Werner developed the coordination theory from a general point of view [3].

Werner's generalizing way of thinking is clearly demonstrated by the fact that he introduced beside the octahedron also the square-planar geometry in his first publication (Figure 1.2). The square geometry is deduced from the observation that compounds $[Ma_2X_2]$ occur in certain cases in two isomeric forms.

It is an interesting point that Werner introduced square-planar coordination geometry with a compound that made history again some 80 years later, as a leading anti-cancer drug [4].

After the first publication, numerous others by Werner followed, where the basic ideas of coordination theory were elaborated in more detail and applied to many concrete cases. The first mention of the theoretical possibility of chiral coordination species appeared in a paper by Werner in 1899 [5]. He calls this type of isomerism 'Spiegelbildisometrie' (Figure 1.3), i.e. mirror image isomerism. Although Lord Kelvin had coined the word 'chirality' already several years earlier[†] [6],

---

[†]In Appendix H of the 'Baltimore Lectures,' by Lord Kelvin [6], the concept of chirality is defined for the first time. This Appendix is the publication of a lecture given by Lord Kelvin in 1893.

Denken wir uns das Metallatom als Zentrum des ganzen Systems, so können wir sechs mit demselben verbundene Moleküle am einfachsten in die Ecken eines Oktaeders verlegen.

Es frägt sich aber, zu welchen Folgerungen diese Annahme führt, und ob diese Folgerungen in den Thatsachen eine Stütze finden.

Denken wir uns zunächst ein Molekül $\left(M_{X}^{(NH_3)_5}\right)$ also in fünf Ecken des Oktaeders Ammoniakmoleküle, im sechsten einen Säurerest.

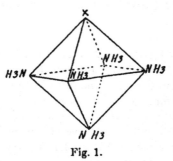

Fig. 1.

Substituieren wir in demselben ein zweites Ammoniakmolekül durch einen Säurerest, so können wir dies auf zwei verschiedene Arten thun.

Entweder können wir das zum Säureradikal axial gelegene Ammoniakmolekül substituieren, oder wir können eines der vier mit ihm an gleichen Kanten des Oktaeders befindlichen Ammoniakmoleküle substituieren, wie folgende Figuren zeigen werden.

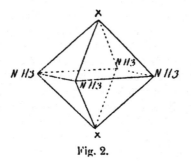

Fig. 2.

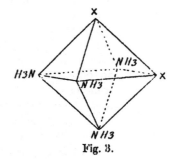

Fig. 3.

Wir müssen also zu zwei isomeren Molekülkomplexen $\left(M_{X_2}^{A_4}\right)$ gelangen.

Figure 1.1
Representation of the first drawings of *octahedra* in chemical literature, from Alfred Werner's fundamental publication in coordination theory [1]

Werner did not use this designation, which came into common usage in chemistry only much later.

As Bernal and Kauffman pointed out [7], Werner had overlooked in his laboratory chiral complex compounds that separate spontaneously upon crystallization into the enantiomeric forms, e.g. $[Co(en)_2(NO_2)_2]Br$ (Figure 1.4), which was prepared by Edith Humphrey in her thesis carried out under the guidance of Werner in 1900. An original sample of $[Co(en)_2(NO_2)_2]Br$ and a CD spectrum of a solution obtained from *one* crystal of that preparation can be seen in the exhibition room of the Royal Society of Chemistry in Burlington House,

$$a\diagdown \diagup x \quad \text{und} \quad x \diagdown \diagup a$$
$$\quad m \qquad\qquad m$$
$$a\diagup \diagdown x \qquad\qquad a\diagup \diagdown x.$$

Nehmen wir z. B. an, die erste Formel entspreche den Plato-semidiamminsalzen, die zweite Formel den Platosamminsalzen

$$Cl\diagdown \diagup NH_3 \qquad\qquad Cl\diagdown \diagup NH_3$$
$$\qquad Pt \qquad\qquad\qquad Pt$$
$$Cl\diagup \diagdown NH_3 \qquad\qquad NH_3 \diagup \diagdown Cl$$

Platosemidiamminchlorid,   Platosamminchlorid.

Sowohl aus Platosemidiamminchlorid, als auch aus Platosemidi-pyridinchlorid mufs nach den oben erwähnten Reaktionen dieselbe Verbindung entstehen.

Figure 1.2
Representation of the first drawings of *square-planar* coordination in chemical literature, from Alfred Werner's fundamental publication in coordination theory [1]

Das unter dieser Voraussetzung sich ergebende Modell ist jedoch, stereochemisch gesprochen, ein asymmetrisches, d. h. es kann in zwei räumlichen Anordnungen, die sich verhalten wie Bild und Spiegelbild und die nicht zur Deckung gebracht werden können, konstruiert werden.

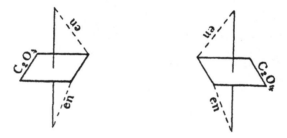

Der gewöhnlichen Asymmetrie bei organischen Molekülen, die bekanntlich die sogenannte optische Isomerie bedingt, ist der ent-wickelte Isomeriefall nicht vergleichbar, weil die hier rechts oder links angeordneten Gruppierungen (2 Äthylendiamine) identisch sind; es würde die obige Isomerie vielmehr vergleichbar sein mit derjenigen organischer Doppelringsysteme, z. B. der folgenden Art:

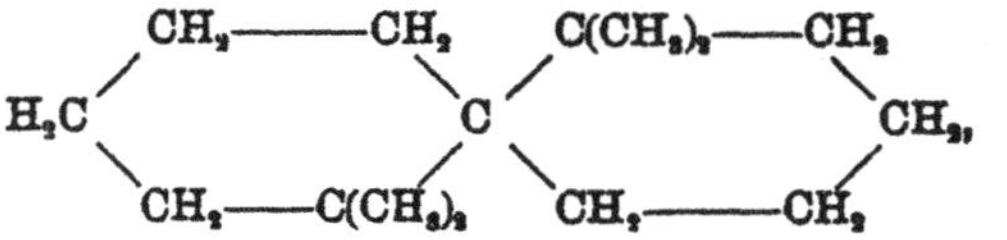

$$CH_2 \text{---} CH_2 \quad C(CH_3)_2 \text{---} CH_2$$
$$H_2C \qquad\qquad C \qquad\qquad CH_2,$$
$$CH_2 \text{---} C(CH_3)_2 \quad CH_2 \text{---} CH_2$$

Figure 1.3
First representation of an *enantiomeric pair* of complexes with octahedral coordination [5]

London (UK), where it was brought as a gift from the Swiss Committee of Chemistry on the occasion of the 150th anniversary of the Society in 1991.

Stereo vision: Figure 1.4 contains two stereo pairs of the two enantiomers of a chiral complex, generated with the computer. Our visual organ perceives through its two eyes two images, which are different with respect to the viewing angle. Our brain constructs from these two images the three-dimensional (3D) world that is familiar to us. With the aid of a pair of lenses, most people can reconstruct from pairs of stereo images a three-dimensional picture of an object, like the model of a molecule. This is highly useful for considerations in stereochemistry, because a

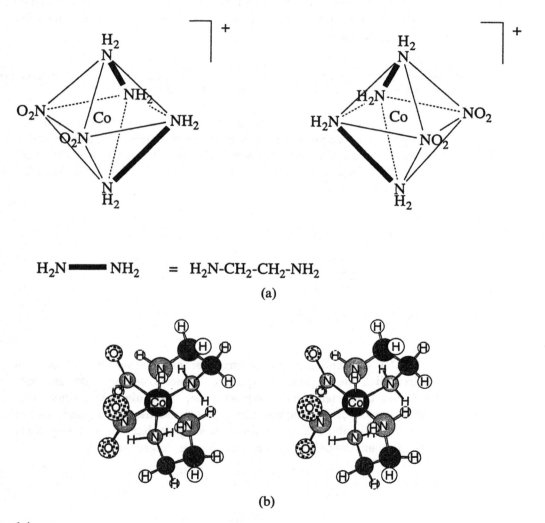

Figure 1.4
Computer generated model of the enantiomeric pair of the chiral complex *cis*[Co(en)$_2$(NO$_2$)$_2$]$^+$. (a) The molecule and its non-congruent mirror image; (b) and (see next page)

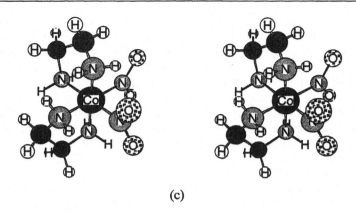

(c)

Figure 1.4 *Continued*
(c) stereo pairs of the two enantiomers

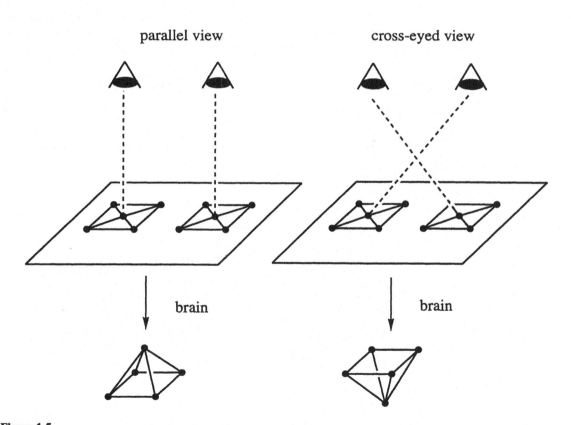

Figure 1.5
Two different possibilities to view stereographs. The brain combines the two single images to a 3D object. Switching from parallel view to cross-eyed view inverts the image. The cross-eyed view can easily be obtained by focusing on a small object (finger, pencil) in front of the picture

purely abstract mental occupation with this subject is often difficult. Methods where one can directly see, in the literal sense, the facts in a *model* are therefore important. Recently, a new method of representing stereo pictures has become popular in graphic arts, the so-called *random dot stereograms* or *autostereograms* [8]. With the random dot stereograms, an unaided three-dimensional vision can be acquired fairly easily. Once the technique of unaided 3D vision has been learned by a person (normally a matter of less than 1 h of initiation plus some subsequent training, in order to become more accustomed to it), it can easily be applied also to stereo pairs of the type shown in Figure 1.4. The random dot stereograms are not indispensable for the learning of this technique, but they facilitate the task considerably.

Two different ways of looking at a stereo pair (and at a random dot stereogram) exist: the 'parallel' and the 'cross-eyed' view (Figure 1.5). Since the left and the right pictures are exchanged in our brain, when one changes from one method to the other an inversion of the picture occurs. For those who have mastered performing the transition from cross-eyed to parallel view (the reverse is more difficult in the author's experience) within seconds, a chiral object changes to its enantiomer in appearance, merely by varying the angle of the eyes. (Of course, stereo vision does not depend on the chirality of an object. Any 3D object can be seen in stereo vision, whether it is chiral or achiral.) The conditions that a stereo pair can be viewed with the parallel *and* the cross eyed techniques are:

- no atom is partially or totally hidden and
- atoms of one kind are of the same size, independent of depth.

An example where these conditions are fulfilled is given in Figure 1.6. The other stereo pairs in the book are all represented for parallel viewing.

It is strongly recommended that people who are seriously interested in stereochemistry try to acquire the skill of the two ways of unaided stereo viewing. In addition, building physical models is very useful for visualizing stereochemical facts.

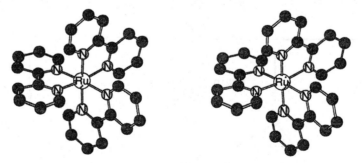

Figure 1.6
Right or left, that is the question! This stereopair of the complex [Ru(bpy)$_3$]$^{2+}$ is a right-handed helix if viewed with the parallel technique, but it changes to a left-handed helix if viewed with the crosseyed technique. Thus both enantiomers of a 3D chiral object can be represented by a single 2D stereo pair, if certain conditions are fulfilled. The result just depends on the relative positions of the two eyes

This compound forms large conglomerate crystals (see Section 4.3.1) of the non-centrosymmetric space group $P2_1$, i.e. it separates spontaneously on crystallization. Werner achieved, together with V. L. King, the first separation of a coordination compound in its enantiomers in 1911 [9,10]. Three years later, the first completely non-organic compound (no carbon present) was also separated by Werner and his co-workers [11].

(1.1)

Even after Werner, many milestones in the development of coordination chemistry were marked by work of stereochemical significance. Russian scientists were among the first to apply Werner's theory. Kurnakov introduced a general reaction to distinguish *cis* from *trans* isomers of Pt^{II} [12,13]. Chugaev had already described the coordination of an asymmetric ligand in 1907 [14]. In 1925 Chernyaev synthesized the three isomers (1.1) of $[Pt(NH_3)(NH_2OH)(py)(NO_2)]^+$ [15], giving additional proof for a square-planar configuration of Pt^{II} and establishing the *trans*-influence series.

The father of coordination chemistry in the USA, John C. Bailar, Jr, was interested in stereochemical problems very early in his career. Bailar and Auten observed the inversion of configuration in reactions of cobalt complexes (see Section 7.1.3.2) [16]. Bailar's preoccupation with stereochemical topics in coordination chemistry lasted for more than 50 years. During this period he published, together with many co-workers, more than 50 papers on stereochemical problems in coordination chemistry [17].

An elegant further proof of square-planar coordination of Pt^{II} complexes was given by Mills and Quibell through the resolution of *meso*-stilbenediaminoisobutylenediaminoplatinum(II) salts [18]. The first, and presumably only, metal analogue to an 'asymmetric' carbon atom, where all ligands are different, the octahedral metal complex $[Pt(Cl)(Br)(I)(NH_3)(NO_2)(py)]$, was synthesized by Essen and Gel'man [19, 20]. Five out of the possible 15 diastereomers were obtained. In (1.2) two different ways of depicting such octahedral complexes are presented.

A particularly important contribution to the field of stereochemistry of coordination compounds came from the Australian school of coordination chemists. It started in the 1930s with the work of D.P. Mellor, and continued with F.P. Dwyer and R.S. Nyholm (who later taught at the University College London) in the 1940s and 1950s [21]. These contributions proved to be highly sustainable, since Australian scientists have been at the forefront of coordination chemistry ever since.

The absolute configurations of the two enantiomers of $[Co(en)_3]^{3+}$ was elucidated shortly after the famous experiment of Bijvoet *et al.* [22], by Saito *et al.* [23]. The

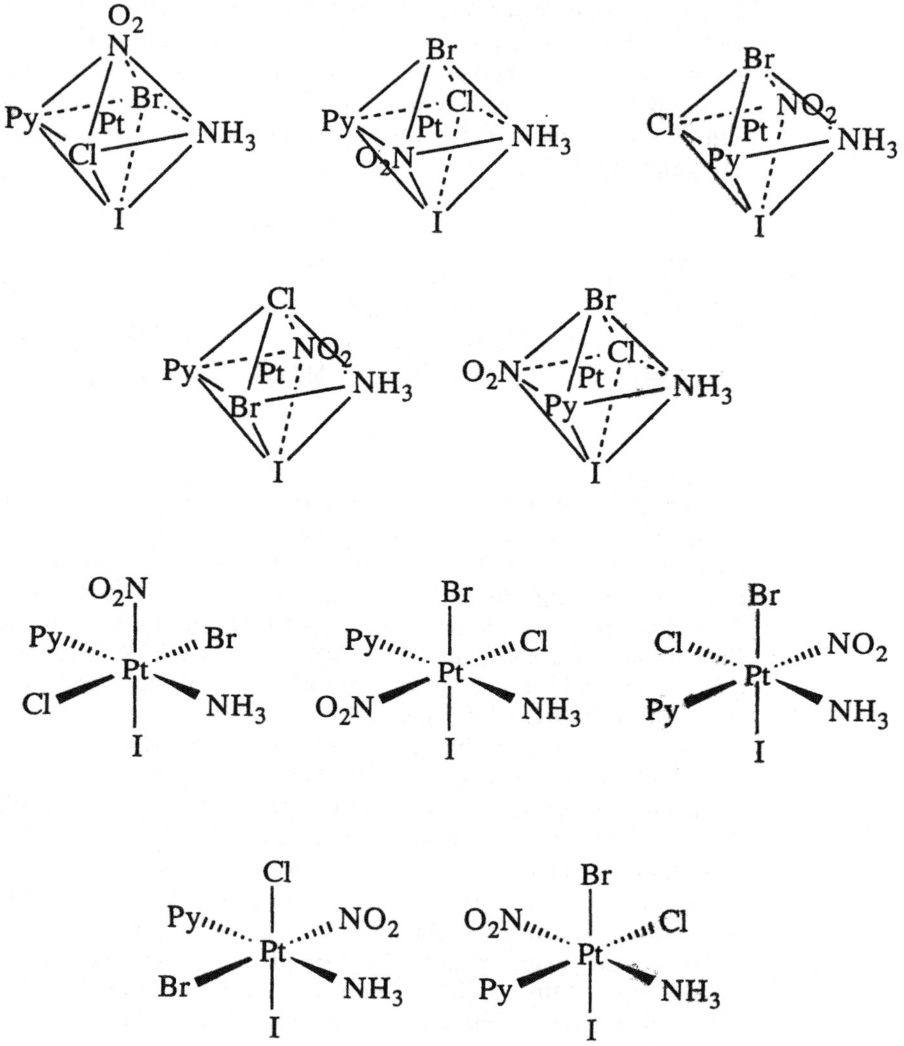

(1.2)

first conformational analysis of coordination compounds was carried out by Corey and Bailar in 1959 [24]. Trigonal prismatic coordination species were discovered by Eisenberg and Ibers in 1965 [25]. Macrocyclic coordination chemistry started with the discovery of cyclam as a ligand [26] and developed into the coordination chemistry of alkali metals with the development of the crown ether ligands [27]. Cage-like ligands were developed on the one hand for alkali metals by Lehn [28], by Boston and Rose [29], and later specifically for transition metals by Sargeson [30,31]. Coordination compounds and their stereochemical behavior are also a central point in the synthesis of catenands and other more intricate structures, such as molecular knots [32–35]. Even comparatively simple coordination compounds exhibiting new stereochemical features can still be found. As an example, square-planar complexes of Pt^{II} with helically twisted ligands were first described in 1990 [36].

In the formulation and in the early history of coordination chemistry, stereochemical considerations were instrumental in the confirmation of the coordination theory of Werner. Later, other more quantitative subjects, such as thermodynamic properties, and a large variety of spectroscopic properties, the electronic structures, catalytic activities, kinetic behavior, etc., became subjects of central interest. Stereochemistry of coordination compounds was often taken for granted and assumed to be essentially understood. Yet coordination geometries with coordination numbers larger than four have an extremely rich stereochemical potential, which is far from a state of complete development. Organic chemists, who deal mainly with coordination number four for carbon, have developed stereochemical descriptions of molecules to a high degree of sophistication. The subject has been treated in classical textbooks [37–39], to which recently a monumental new edition has been added [40]. It can be stated that a difference exists between organic and inorganic stereochemistry, inasmuch as carbon is unique among the elements in the periodic table, in forming a large number (in fact, the number of possible organic molecules is not limited) of *substitution inert*[†] molecules with tetrahedral coordination. Although a number of metallic elements exhibit tetrahedral coordination, few of them are really substitutionally inert. Most substitutionally inert metal complexes have either square-planar or octahedral coordination spheres.

A field where the stereochemistry of molecules has a bewildering variability is organometallic chemistry. The capability of metallic elements to form multi-center bonds, either in π-complexes as in the metallocenes, or in cluster compounds, brings a new dimension to their structural properties. Even though stereochemical concepts should not be restricted to certain areas of chemistry, it is difficult to include a large class of organometallic compounds into the general discussion of the stereochemistry of coordination compounds. The concepts treated in this book should be applicable to organometallic compounds, but they will often not suffice to describe their stereochemical properties completely. The treatment is restricted to Werner-type compounds, where in general one or several coordination centers having an unambiguously definable coordination number are clearly recognizable. This includes some organometallic species, e.g. some cyclometalated complexes, and the 'tetrahedral' complexes with three electron pair donor ligands and one π-bonded cyclic ligand of the cyclopenetadienyl (cp) type, but it excludes many other molecules, especially all those with metal–metal bonds, i.e. the so called cluster compounds.

Recently, renewed interest in the stereochemistry of coordination compounds has arisen, and many important new results have been obtained in this field. Several developments have contributed to that renewal of interest, among them most importantly: *enantioselective catalysis* [41,42], *bioinorganic chemistry* [43,44], and *supramolecular chemistry* [45]. The strong interest in these fields combined with the

---

[†]An atomic center in a molecule is called substitution inert if the lifetime of its first coordination sphere under given experimental conditions is large on the time-scale of separation procedures.

availability of highly informative experimental methods, especially X-ray diffraction and NMR spectroscopy, will undoubtedly put the stereochemistry of coordination compounds into the class of pre-eminent topics in the science of the 21st century.

The early history of coordination chemistry, which is essentially a history of stereochemistry of coordination compounds, has been reviewed in great detail in many papers and books by Kauffman [46].

1.2   Basic Concepts: General Considerations about Stereochemistry, Structure, Geometry, and Symmetry. Metric and Topographic Stereochemistry

It is interesting to note that stereochemistry addresses different problems in different fields of chemistry. It *is* strongly related to questions of *stereoisomerism* in most treatments of organic molecules [37,40], and it *was* the same for coordination compound in Werner's time. This aspect, however, is not considered at all in some contemporary books on inorganic stereochemistry [47,48]. In both of these books the emphasis is on quantitative aspects (coordination geometries, distances, bond angles) of molecular geometry, as expressed nicely in the title of Burdett's book *Molecular Shapes*. Both viewpoints are evidently important for a description of molecules and both have therefore their *raison d'être*. We propose to distinguish the two branches of stereochemistry by naming them *topographic* [49] and *metric stereochemistry*, respectively. In the present treatment both viewpoints will be considered, emphasizing, however, the topographic aspects, i.e. stereoisomerism and connectivity in molecules. The metric aspects will be considered only as far as it is necessary as a basis for discussions of topographic stereochemistry of coordination compounds.

The theory of *metric stereochemistry* is to a large extent the theory of chemical bonding. Two slightly overlapping methods are used today for theories of chemical bonding: *Computational methods* make use of the ability of modern computers to approximate solutions of the basic equations of Quantum Mechanics, whereas *'chemical theories'* derive concepts which are based on quantum chemical reasoning, but which take into account certain facts of descriptive chemistry. Several such theories can be applied to the problems of metric stereochemistry. Burdett's treatment of inorganic molecules, including a great number of coordination compounds, is strongly recommended for the reader who wants to delve more deeply into this field. These 'chemical' theories (VSEPR, CF, MO, AOM, etc.) provide some general concepts which can be applied to whole classes of compounds in a useful way. Computational theory on the other hand, can provide some interesting results for systems that are not too complicated.

Topographic stereochemistry can be divided into three classes of problems (which are interconnected): *symmetry, topology,* and *enumerational properties*. At first sight it seems that all of them should be tractable by mathematics. Indeed, for all these subjects, mathematical methods exist which can be applied to chemical problems.

Chemists, in particular inorganic chemists, have become used to the application of *group theory* to problems dealing with symmetry. There is little doubt that group theory is very useful for the purpose of classification of objects according to their symmetry, and also for the derivation of certain molecular properties, such as the determination of degeneracies of electronic states and assigning vibrational modes.

The application of basic concepts of *topology* in stereochemistry has been discussed by several authors [50–52]. In general, chemists do not need the heavy artillery of mathematical topology, but basic considerations about topological aspects of molecules can yield interesting viewpoints.

Combinatorial enumerations, e.g. the often not highly complicated problem of the number of isomers, and dynamic problems are, in principle, also amenable by mathematics [53]. Yet the formal description becomes so involved [54–56], [57, p.216] that it is, at least in actual cases often not very practical to apply these methods. This book is written for the experimental chemist, and not for the mathematical chemist. The chemist who has to solve a stereochemical problem, such as finding the number of possible isomers of a molecule, will generally not go to the trouble of using abstruse mathematical methods. As demonstrated by many fundamental concepts in chemistry and in many other fields (e.g. the genetic code as a basis for all genetic information), *understanding* and *solving* scientific problems by no means always signify the necessity to put a theory into an abstract mathematical form. It will be useful to put mathematics to work in order to develop some concepts of stereochemistry further, but the experimental chemist will always find well defined simple concepts more useful than mathematical derivations.

It is important, however, to define the concepts that are used as clearly as possible, and in accordance with a theoretically sound basis. An example of a concept which had been used for more than 100 years in a seemingly well defined way, but which had to be reformulated on a different basis, is the relationship between an 'asymmetric' carbon atom and the chirality of the molecule. Since the time of van't Hoff and Le Bel, the discoverers of the tetrahedral carbon atom, organic chemists struggled with so-called 'pseudo-asymmetric' carbon atom (1.3) (see also Section 4.2.2). The pseudo-asymmetric carbon atom has four different ligands, but lies on a plane of symmetry. In a paper in 1984, Mislow and Siegel [58] showed that the asymmetric carbon atom is a concept which leads, at least in connection with its relation to chirality, to paradoxical situations such as 'pseudo-asymmetry'.

$$a$$
$$b \qquad C_R(x,y,z)$$
$$C_S(x,y,z)$$

(1.3)

Topographic stereochemistry is ubiquitous in all parts of chemistry. Its concepts were, in the past, most vigorously developed by organic chemists, yet the stupendous stereochemical variability of molecules comprising atoms with coordination numbers greater than four calls for a comparable effort to be made by coordination chemists.

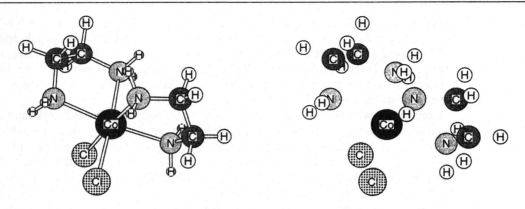

Figure 1.7
Computergenerated models of [CoCl$_2$(en)$_2$]$^+$, with and without connectivities indicated between the atoms

Most considerations in stereochemistry rely strongly on the concepts of *molecules* and their *structures*. Both these concepts are for every chemist intuitively quite clear in their meaning, yet both are not easy to define in a qualitatively unique way. Important questions are, of course: What is a chemical bond? How many kJmol^{-1} signifies a bond? Where is the limit to a hydrogen bridge or another interaction generally designated as intermolecular? Is a *polymer* an *isomer* of the monomeric unity? For all chemistry the *connectivity of atoms* in any given molecular arrangement is an extremely important way of visualizing the molecular structure. It is an interesting experience to construct a molecule which is not even very complicated, e.g. *[Co(en)$_2$Cl$_2$]$^+$*, with a computer program on screen in a ball and stick representation and than to take away all the symbols for connectivity, i.e. the sticks. Even an experienced coordination chemist will find it difficult to recognize the structure of the molecule (Figure 1.7). The rules for setting up the connectivity in a molecule are usually quite obvious, yet they are by no means laws of nature. Problems of setting up the connectivity arise especially in *organometallic* and *cluster chemistry*. Other, seemingly clear concepts, which are essential for stereochemistry are also not so simple to define in a strict sense. Take for example the distinction between isomer and *excited state*. Is an electronically excited state to be considered an isomer with respect to the ground state? How long must an atomic arrangement persist, such that the structure can be called a molecule? Later in this book, we will try to elucidate as clearly as possible the different kinds of isomers which exist for a given structure, but we will not try to define fundamental concepts, like the general meaning of *molecule*, *structure* or *isomer* in a way to satisfy the logical purist. Problems with other concepts may occur, because a phenomenon or behavior has been given a name, which is not very meaningful or even misleading. For example *optical activity*[†] means something (more or less) well defined in chemistry, but from a general point of view it is not a very informative

---

[†]Every chemist knows, at least qualitatively, what phenomenon is described by the concept of optical activity. The concept is, however, neither widely known among other scientists, nor self-explanatory.

name. In addition, 'optical activity' is often used in a slack way.‡ The same is true, even to a larger extent, for the concept of *geometrical isomers*. The latter is still used frequently by coordination chemists, despite the fact that is was long ago recognized to be a misnomer [59].

1.3 References

1. Werner, A. (1893), *Z. Anorg. Chem.*, **3**, 267–330.
2. Werner, A., *Lehrbuch der Stereochemie*, Verlag von Fischer G., Jena, 1904, pp. 317–350.
3. Werner, A., *Neuere Anschauungen auf dem Gebiete der Anorganischen Chemie*, 3. Auflage, F. Vieweg & Sohn, Braunschweig, 1913.
4. Rosenberg, B., VanCamp, L., Trosko, J. E. and Mansour, V. H. (1969), *Nature*, **222**, 385–386.
5. Werner, A. and Vilmos, A. (1899), *Z. Anorg. Allg. Chem.*, **21**, 145–164.
6. Lord Kelvin, in *Baltimore Lectures*, Cambridge University Press, Cambridge, 1904.
7. Bernal, I. and Kauffman, G. B. (1993), *Struct. Chem.*, **4**, 131–138.
8. Thing Enterprises, N. E., *The Magic Eye: A New Way of Looking at the World*, Andrews and McMeel, Kansas City, 1993.
9. King, V. L. (1942), *J. Chem. Educ.*, **19**, 345.
10. Werner, A. (1911), *Chem. Ber.*, **44**, 1887–1890.
11. Werner, A. (1914), *Chem. Ber.*, **47**, 3087–3094.
12. Kauffman, G. B. and Beck, A. (1962), *J. Chem. Educ.*, **39**, 44–49.
13. Kurnakov, N. S. (1893), *J. Russ. Phys. Chem. [2]*, **50**, 481–507.
14. Chugaev, L. (1907), *J. Prakt. Chem. [2]*, **76**, 88–93.
15. Chernyaev, I. I. (1926), *Izv. Inst. Izuch. Plat. Drugikh Blagorodn. Metal*, **4**, 243–275.
16. Bailar, J. C., Jr, and Auten, R. W. (1934), *J. Am. Chem. Soc.*, **56**, 774–776.
17. Bailar, J. C., Jr (1990), *Coord. Chem. Rev.*, **100**, 1–27.
18. Mills, W. H. and Quibell, T. H. H. (1935), *J. Chem. Soc.*, 839–846.
19. Essen, L. N. and Gel'man, A. D. (1956), *Zh. Neorg. Khim.*, **1**, 2475.
20. Essen, L. N., Zakharova, F. A. and Gel'man, A. D. (1958), *Zh. Neorg. Khim.*, **3**, 2654–2661.
21. Livingstone, S. E., in *The Contributions of David P. Mellor, Frank P. Dwyer, and Ronald S. Nyholm to Coordination Chemistry*, G. B. Kauffman (Ed.), *ACS Symposium Series, No. 565: Coordination Chemistry: A Century of Progress 1893–1993*, American Chemical Society, Washington, DC, 1994, Chap.10, pp.126–135.
22. Bijvoet, J. M., Peerdeman, A. F. and Van Bommel, A. J. (1951), *Nature*, **168**, 271–272.
23. Saito, Y., Nakatsu, K., Shiro, M. and Kuroya, H. (1955), *Acta Crystallogr.*, **8**, 729–730.
24. Corey, E. J. and Bailar, J. C., Jr (1959), *J. Am. Chem. Soc.*, **81**, 2620–2629.
25. Eisenberg, R. and Ibers, J. A. (1965), *J. Am. Chem. Soc.*, **87**, 3776–3778.
26. Curtis, N. F. (1960), *J. Chem. Soc.*, 4409–4413.
27. Pedersen, C. J. (1967), *J. Am. Chem. Soc.*, **89**, 7017–7036.
28. Lehn, J. M. (1988), *Angew. Chem., Int. Ed. Engl.*, **27**, 89–112.
29. Boston, D. R. and Rose, N. J. (1968), *J. Am. Chem. Soc.*, **90**, 6859–6860.
30. Sargeson, A. M. (1979), *Chem. Br.*, **15**, 23–27.
31. Sargeson, A. M. (1991), *Chem. Aust.*, **58**, 176–178.
32. Chambron, J. C., Dietrich-Buchecker, C. and Sauvage, J.-P. (1993), *Top. Curr. Chem.*, **165**, 131–162.
33. Dietrich-Buchecker, C. O., Guilhem, J., Pascard, C. and Sauvage, J.-P. (1990), *Angew. Chem., Int. Ed. Engl.*, **29**, 1154–1156.

---

‡One can often read that an 'optically active molecule' (instead of a 'chiral molecule') is obtained in a synthesis, even if the compound synthesized is a racemate. In addition, optical activity is generally not observed with single *molecules*, but only in solutions that contain an excess of one of the two forms of a chiral molecule.

34. Dietrich-Buchecker, C. O. and Sauvage, J.-P. (1987), *Chem. Rev.*, **87**, 795–810.
35. Sauvage, J.-P. (1985), *Nouv. J. Chim.*, **9**, 299–310.
36. Deuschel-Cornioley, C., Stoeckli-Evans, H. and Von Zelewsky, A. (1990), *J. Chem. Soc., Chem. Commun.*, 121–122.
37. Eliel, E. L., *Stereochemistry of Carbon Compounds*, 1st edn, McGraw-Hill, New York, 1962.
38. Potapov, P. A., *Stereochemistry*, Mir, Moscow, 1979.
39. Testa, B., in *Principles of Organic Stereochemistry, Studies in Organic Chemistry*, Vol. 6, P. G. Gassman (Ed.), Marcel Dekker, New York, 1979.
40. Eliel, E. L. and Wilen, S. H., *Stereochemistry of Organic Compounds*, Wiley-Interscience, New York, 1994.
41. Brunner, H. and Zettlmeier, W., *Handbook of Enantioselective Catalysis with Transition Metal Complexes Ligands – References*, Vol. II, VCH, Weinheim, 1993.
42. Brunner, H. and Zettlmeier, W., *Handbook of Enantioselective Catalysis with Transition Metal Complexes – Products and Catalysts*, Vol. I, VCH, Weinheim, 1993.
43. Frausto da Silva, J. J. R. and Williams, R. J. P., *The Biological Chemistry of the Elements – The Inorganic Chemistry of Life*, Clarendon Press, Oxford, 1993.
44. Kaim, W. and Schwederski, B., *Bioinorganic Chemistry: Inorganic Elements in the Chemistry of Life. An Introduction and Guide*, Wiley, Chichester, 1994.
45. Balzani, V. and De Cola, L. (Eds), *Supramolecular Chemistry, NATO ASI Series*, Kluwer, Dordrecht, 1992.
46. Kauffman, G. B., *Inorganic Coordination Compounds. Nobel Prize Topics in Chemistry, A Series of Historical Monographs on Fundamentals of Chemistry*, Heyden, London, 1981.
47. Burdett, J. K., *Molecular Shapes – Theoretical Models of Inorganic Stereochemistry*, Wiley, New York, 1980.
48. Kepert, D. L., *Inorganic Stereochemistry, Inorganic Chemistry Concepts*, Vol. 6, Springer, Berlin, 1982.
49. Walba, D. M., in *Chemical Applications of Topology and Graph Theory*, R. B. King (Ed.), Elsevier, New York, 1983.
50. Frisch, H. L. and Wassermann, E. (1961), *J. Am. Chem. Soc.*, **83**, 3789–3795.
51. King, R. B. (1991), *J. Math. Chem.*, **7**, 51–68.
52. Sauvage, J.-P. (Ed.), *Topology in Molecular Chemistry*, in *New J. Chem.*, 1993, **17**, 617–757.
53. Ugi, I., Dugundij, J., Kopp, R. and Marquarding, D., *Perspectives in Theoretical Stereochemistry*, Springer, Berlin, 1984.
54. Polya, G. and Read, R. C., *Combinatorial Enumeration of Groups, Graphs and Chemical Compounds*, Springer, New York, 1987.
55. Schumacher, E. (1994), *Chimia*, **48**, 26–29.
56. Shinsaku, F., *Symmetry and Combinatorial Enumeration in Chemistry*, Springer, Berlin, 1991.
57. Sokolov, V. I., *Introduction to Theoretical Stereochemistry*, translated from Russian by Standen, N. F., Gordon and Breach, New York, 1991.
58. Mislow, K. and Siegel, J. (1984), *J. Am. Chem. Soc.*, **106**, 3319–3328.
59. Prelog, V., in *Van't Hoff–Le Bel Centennial*, B. Ramsey (Ed.), *ACS Symposium Series*, Vol. 12, American Chemical Society, Washington, DC, 1975, pp. 179–188.

2   Survey of Methods for the Elucidation of the Stereochemistry of Coordination Compounds

The study of the stereochemistry of molecular species requires methods to study the arrangements of atoms in three-dimensional space. Since atomic dimensions are below a scale where a direct linkage between molecular structures and the macroscopic world of our daily experience can be established[†], such methods are indirect. Until about 1950, most methods which could elucidate the topographic stereochemistry of molecules were in fact largely chemical in nature. It was the number and the kind of isomers or in some cases the reactivity, such as the *trans*-influence experiments with square-planar complexes, that gave the clues for the molecular structure. Although X-ray diffraction methods were invented and theoretically understood much earlier, it was not until the middle of the 20th century that they became applicable to molecules of a certain complexity. The possibility of solving the crystal and the molecular structure of many chemical compounds was one of the first and most important impacts computers had on the development of chemistry in general, and of stereochemistry in particular. The early reports about application of the X-ray method by Wyckoff and Posnjak [1] and Dickinson [2] proved in several cases ($[PtCl_6]^{2-}$, $[PdBr_6]^{2-}$, $[SnCl_6]^{2-}$, $[PtCl_4]^{2-}$, $[PdCl_4]^{2-}$, $[Co(NH_3)_6]^{3+}$, $[Ni(NH_3)_6]^{3+}$) octahedral and square-planar coordination directly. They were considered to be triumphs for both coordination theory and for the diffraction method.

Two physical methods, which were available much earlier, were polarimetry (i.e. the measurement of the rotation of linearly polarized light) and dipole moment determinations. They contributed enormously to the development of organic and inorganic stereochemistry. Werner, the discoverer of chiral coordination compounds, used polarimetry at various wavelengths [optical rotatory dispersion (ORD)] to a large extent. This was already possible at the beginning of the century since optics was a technology that was developed relatively early.

Other, more powerful methods had to await the advent of modern electronics and computational methods, which occurred in the second half of the century. Today there exist a variety of methods that are very useful for the elucidation of molecular structure. These methods can in general be divided into two groups:

[†]This statement has to be relativized, since with STM (scanning tunnel microscopy) and AFM (atomic force microscopy) methods, objects of the size of single atoms can be visualized in a direct way. Even in the future, those methods will be, however, more apt to the elucidation of structural properties of assemblies of molecules or of the higher order structures of macromolecules, rather than for the investigation of the stereochemistry of relative small molecules.

Diffraction methods and *spectroscopic methods*. In each of the two groups, there exists one method whose application is far more prevalent than others. In the former it is *X-ray diffraction*, and in the latter *NMR spectroscopy*. They are complementary in the sense that X-ray diffraction is carried out mainly with crystalline substances, whereas NMR spectroscopy is most suitable for species in solution. In addition, X-ray diffraction yields complete structural information, i.e. atomic coordinates, whereas NMR spectroscopy has been, and still is today, mostly used as a method yielding information on topographic stereochemistry only. The high popularity of these two methods does not mean that they are superior to others for *all* structural problems. Some systems require the use of other methods, such as neutron diffraction, UV-visible spectrophotometry, vibrational spectroscopy, EPR spectroscopy, Mössbauer-spectroscopy, etc. A comprehensive presentation and discussion of the methods most frequently used in structural inorganic chemistry is given in a recent book [3].

Diffraction methods are the only methods that give direct information on the positions of atoms in three-dimensional space. It is left to the chemist (or to the crystallographer) to set up the connectivity in a molecule. X-ray diffraction yields a mapping of the electron density of molecules that can be directly related to the positions of atoms. With the methods generally applied to resolve the structure, only the relative configuration of a molecule can be determined, i.e. a distinction of a structure and its mirror image is not possible. In the case of chiral molecules, where the molecule and its mirror image are not identical, the method of Bijvoet *et al.* [4,5] (anomalous diffraction of X-rays) can remove this ambiguity, and a determination of the absolute configuration becomes possible. Nowadays, the absolute configurations of so many enantiomers are known that a determination of a relative configuration of a molecule often gives its absolute configuration through connection with some structure of known absolute configuration. In addition, high quality diffraction data allow for a routine application of Bijvoet *et al.*'s anomalous scattering method.

Despite, or perhaps just because of the highly appealing features of the X-ray diffraction method, which yields information corresponding to the ways chemists think of molecular structures, a caveat is appropriate. X-ray diffraction is a sampling method, where the sample examined is a small fraction of the total amount of a substance prepared or isolated and, in most cases, the sample has been specially prepared by a phase transition process (crystallization). The diffraction data are collected using *one* crystal of a volume of typically $10^{-11}m^3$, having a mass of typically about $10\mu g$, corresponding to about 10nmol. This is normally a tiny fraction of the whole preparation. Even if the compound is characterized by other methods, it is by no means certain that all the material prepared corresponds to the structure determined by the diffraction measurement. Coordination compounds can often give a number of different isomers that can only be distinguished by structure-sensitive methods, and not by ordinary analytical procedures. It is not known how many structures are published in chemical literature that represent only a fraction of the substance under consideration.

Although NMR is used in most cases for topographic information only, it can yield relative positional parameters of nuclei through experiments, which rely on well established interactions between nuclei, such as magnetic dipole–dipole interactions. These methods have been applied successfully for structures of organic or bioorganic molecules in solution [6], and also for coordination compounds [7,8]. In most cases, topographic stereochemical information is obtained through symmetry properties, which often show up in NMR spectra in a surprisingly simple and unambiguous way. There is no direct analogy to X-ray diffraction in NMR spectroscopy for the elucidation of absolute configuration. If, however, a chiral molecule is brought into a non-racemic environment, e.g. by adding suitable enantiomerically pure chiral shift reagents, some information about absolute configurations can also be gained through NMR spectroscopy. Most other structural methods do not give information about positional parameters of atoms in molecules, or about absolute configurations, with the exception of chiroptical methods (circular dichroism and optical rotatory dispersion), which give indications about relative and, through theoretical or analogy considerations, also about absolute configurations [9–11].

As a bulk method, which operates with solution samples that represent the whole preparation of a substance, NMR does not have the problem of a possible determination of a non-representative structure, as does X-ray diffraction. NMR usually yields information about structure, isomeric purity, and purity in general. Therefore, for complete structural information, X-ray diffraction *and* NMR investigations should be carried out whenever they are feasible.

All methods for structural investigations of molecules, diffraction as well as spectroscopic methods, rely on the interaction of electromagnetic radiation with matter. In such interactions time is a very important parameter for the information sought after. Through Planck's relation and the uncertainty principle,

$$E = h\nu$$

$$\Delta E \cdot \Delta t \geqslant \hbar$$

a non-quantum mechanical' expression (not containing Planck's constant h)

$$\Delta \nu \cdot \Delta t \geqslant 1$$

is obtained. Therefore, methods using high-frequency radiation, in particular X-ray diffraction, give essentially instantaneous structural information, even on the scale of molecular vibrations. The same is true for UV-visible spectrophotometry. Since such experiments are carried out with macroscopic samples, i.e. with a large number of molecules, ensemble distributions of structural parameters are obtained. This yields, e.g. in the case of X-ray diffraction, important information about mean vibrational amplitudes of atoms in molecules. Although the so-called thermal parameters due to vibrational displacements are seldom analyzed in detail, they can contain, in principle, interesting stereochemical information [12].

With low-frequency methods, especially NMR and EPR spectroscopy (with NMR more often than with EPR), time-averaged information is obtained for structural changes that are fast compared with the frequency resolution of the method. Well known examples are NMR spectra of five-coordinated complexes ML_5, where nuclei in the ligands L often show up as completely equivalent [13], although no coordination sphere exists where five ligands can be equivalent from a geometric point of view. The frequency resolution of NMR spectroscopy is typically of the order of 10–100Hz. Therefore, fast or even not so very fast processes, e.g. conformational changes or certain types of isomerizations, may yield completely time-averaged structural information. In many cases, such processes will be just in the rate domain where they cause relaxation effects, observable through specific line broadening in the spectra. If this is the case, kinetic information about stereochemical changes in molecules can be obtained directly from such spectra. The same observations are possible in EPR spectroscopy, except that the processes which can cause line broadening are much faster, in accordance with the lower frequency resolution (generally $\Delta v > 1\text{MHz}$) of the method.

Finally, theoretical methods can be applied nowadays to predict the stereochemistry of molecules in certain special circumstances. There are some cases known where theoretical methods have given the correct structure of molecules before experimental results were available (cf. Chapter 5; TP-6 vs OC-6 CG in d^0 complexes). In general, however, it must be stated that the modern *ab initio* calculational methods are still not necessarily very reliable, since interactions such as solvation are very difficult to treat in this way. Chemistry in general, and stereochemistry in particular, is to a large degree an experimental science and it will remain so for a long time to come. However, computational methods will become more and more important as sources of additional and supportive information.

In summarizing this short survey of structural methods, it can be stated that coordination chemistry always requires the use of structural methods. The two methods emphasized are especially versatile: topographic stereochemical information is often easily accessible (in diamagnetic systems) through NMR spectroscopy. X-ray diffraction yields both types of information, metric and topographic, but it requires crystals and its application is more involved than NMR spectroscopy. UV-visible and IR spectroscopy are also routinely applied in order to obtain some stereochemical information. More specialized methods such as EPR, NQR and Mössbauer [14] spectroscopy should be considered for certain cases, as is the case with the chiroptical methods.

2.1   References

1. Wyckoff, R. W. G. and Posnjak, E. (1921), *J. Am. Chem. Soc.*, **43**, 2292–2309.
2. Dickinson, R. G. (1922), *J. Am. Chem. Soc.*, **44**, 2404–2411.
3. Ebsworth, E. A. V., Rankin, D. W. H. and Cradock, S., *Structural Methods in Inorganic Chemistry*, 2nd edn, Blackwell, Oxford, 1991.
4. Bijvoet, J. M., Peerdeman, A. F. and Van Bommel, A. J. (1951), *Nature*, **168**, 271–272.
5. Eliel, E. L. and Wilen, S. H., *Stereochemistry of Organic Compounds*, Wiley-Interscience, New York, 1994.
6. Wüthrich, K., *NMR of Proteins and Nucleic Acids*, Wiley, New York, 1986.
7. Albinati, A., Isaia, F., Kaufmann, W., Sorato, C. and Venanzi, L. M. (1989), *Inorg. Chem.*, **28**, 1112–1122.
8. Haelg, W. J., Oehrstroem, L. R., Ruegger, H. and Venanzi, L. M. (1993), *Magn. Reson. Chem.*, **31**, 677–684.
9. Jensen, H. P. and Woldbye, F. (1979), *Coord. Chem. Rev.*, **29**, 213–235.
10. Mason, S. F., in *Fundamental Aspects and Recent Developments in Optical Rotatory Dispersion and Circular Dichroism*, P. Salvadori and F. Ciardelli (Eds), Proc. NATO Adv. Study Inst., Heyden, London, 1973, pp. 196–239.

11. O'Brien, T. D., McReynolds, J. P. and Bailar, J. C., Jr (1948), *J. Am. Chem. Soc.*, **70**, 749–755.
12. Dunitz, J. D., Maverick, E. F. and Trueblood, K. N. (1988), *Angew. Chem., Int. Ed. Engl.*, **27**, 880–895.
13. Muetterties, E. L. (1970), *Acc. Chem. Res.*, **3**, 266–273.
14. Gütlich, P., Link, R. and Trautwein, A., *Mössbauer Spectroscopy and Transition Metal Chemistry*, Springer, Berlin, 1978.

3 Theoretical Bases for the Coordination Geometries of Metal Ions

3.1 Coordination Geometries

This chapter gives a cursory view of *metric stereochemistry*, as it is needed as a basis for the discussion of topographic stereochemistry. The reader is referred to other texts quoted in this chapter for a more quantitative consideration of bonding parameters in metal complexes.

A 'central' atom is central in the sense that we *define* this atom to be central.[†] As we consider it and its immediate surrounding, it has in general, in most known molecules, a fairly well defined first coordination sphere. That means that a certain number of adjacent atomic centers can be unambiguously distinguished from other atomic centers assumed to be not directly linked to the central atom under consideration. This number is called the coordination number (CN) of the atomic center. The adjacent atoms are called ligand atoms. If they are covalently bound to other atoms, the whole molecular unit (charged or uncharged) is called a ligand (see Appendix III). The ligands are usually still called ligands, even if they are detached (in most cases imagined as a consequence of a heterolytic bond cleavage) from the 'central' atom. To distinguish the two cases more clearly, one can speak of a 'coordinated ligand,' or a 'free ligand.'[‡]

The atoms defining the CN generally define a geometric arrangement that, in simple cases, corresponds to a geometrically precise polyhedron, whereas in others it is approximating more or less closely a polyhedron. As was shown by Muetterties and Guggenberger [3], coordination species represent or most often approach idealized polyhedra, which are *fully triangulated*, although clearly cases occur where some faces are not triangles. It was proposed by these authors that for the discussion of the metric stereochemistry of coordination species, the dihedral angles formed by the normals to adjacent polytopal faces are given as a quantitative measure of shape. A distinction exists between coordination geometry (CG) and the coordination symmetry (CS). The former is the polyhedron, most

---

[†]The atom may be centrally located in the molecule in the sense that it lies exactly or approximately on the point of the symmetry point group (or the center of mass) of the molecule, as is the case, e.g. in most mononuclear coordination complexes, but it can also lie in any part of a molecule

[‡]The use of 'ligand' in the sense explained above has recently been severely criticized [1]. Since its introduction by A. Stock [2], the usage of this word has clearly evolved in the sense we propose and it would be extremely difficult to introduce a new, cumbersome expression, just for the sake of linguistic purism. We do not feel that we lose a word, as Professor G. J. Leigh fears, because the so-called 'slack usage' has proved to be very useful many millions of times. Professor Leight proposes to use the word 'proligand' for a 'free ligand'. There are numerous other examples of words that have evolved from their original strict meaning. We have never felt, for instance, the need to call the lid of a cooking pot a 'prolid', because it was, for an instant, not covering the pot.

closely approximated by the ligand atoms in the sense of Muetterties and Guggenberger, whereas the latter indicates the actual symmetry (described by the point group) of the molecular skeleton, where the atomic nuclei (called atoms for short) are at their positions of highest probability if molecular vibrations are taken into account. As an example, both types of complexes $[Ma_6]$ and $[Ma_4b_2]$, where a and b are monoatomic ligands, show in general octahedral CG, but $[Ma_4b_2]$ cannot have octahedral CS. $[Ma_6]$ *may have* octahedral CS, but it may also be 'distorted'.[†] It can deviate from exact octahedral symmetry, e.g. by a phenomenon called the Jahn–Teller Effect (in Cu^{2+} complexes), by the influence of a non-bonding electron pair (in $[XeF_6]$), or by effects of the environment. One coordination geometry can, in general, be obtained from another one (having the same CN) by a continuous transformation of relative atomic positions.

A coordination center is called *homoleptic* (Figure 3.1) if all coordinated *ligands* are chemically equivalent, otherwise it is called *heteroleptic*. Homoleptic and heteroleptic are defined *with respect to ligands*, and *not* with respect to ligand atoms. The number of different ligands can be specified by a prefix (bis-, tris-heteroleptic, etc.). Phosphorus in $[PF_5]$ is a homoleptic coordination center, although the five fluorine ligands cannot, for geometric reasons be equivalent (no polyhedron with five equivalent vertices exists). On the other hand, P in $[PCl_3F_2]$ is a bis-heteroleptic coordination center.

From a geometric point of view a large number of arrangements are possible, even for the lower CNs [5–9], but there is a limited number of CGs, which have to be considered in chemistry in general. As mentioned earlier, these CGs often correspond to the fully triangulated polyhedra [3]. It is very important for practical stereochemical purposes that only a restricted number of polyhedra have to be

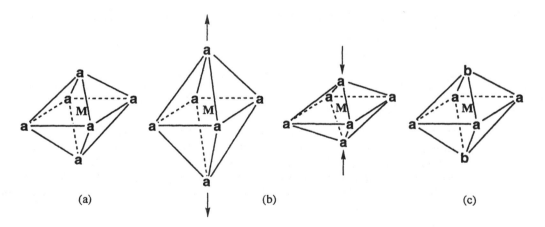

(a)   (b)   (c)

Figure 3.1
(a) $[Ma_6]$, homoleptic, CN = 6; coordination geometry (CG), octahedral; coordination symmetry (CS), octahedral (O_h); (b) $[Ma_6]$, homoleptic, CN = 6; CG, octahedral; CS, tetragonal (D_{4h}); (c) $[Ma_4b_2]$, bisheteroleptic, CN = 6; CG, octahedral; CS, tetragonal (D_{4h}).

[†]The reduction in symmetry, by removing symmetry elements from a higher symmetry structure, has been called *desymmetrization* (Ref. 4, p. 88).

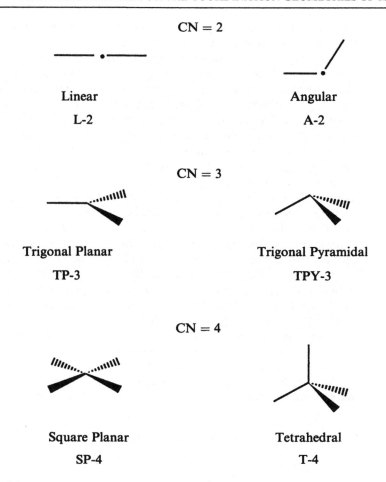

Figure 3.2
Representation of stereo centers for CN $\leqslant$ 8. [†]TBPY-5 in the IUPAC recommendations

considered. The International Union of Pure and Applied Chemistry (IUPAC)[†] Commission on Nomenclature for inorganic compounds has proposed descriptors for these frequently encountered polyhedra. The detailed prescriptions concerning stereochemical nomenclature will be discussed later in connection with isomerism of coordination compounds. Here the descriptors for CGs are introduced, since they will be used throughout the text.

These CG descriptors are called polyhedral symbols in the official IUPAC nomenclature [10], the *Red Book*.[‡] They are straightforward, one-, two- or three-letter symbols for the geometry followed by a dash followed by the CN. For CNs

---

[†]International Union of Pure and Applied Chemistry
[‡]In other publications [11], and also (sometimes only) in the IUPAC *Red Book* these symbols are called 'site symmetry symbols,' which is a misleading term since the site symmetry is often different from the full symmetry represented by the idealized polyhedron.

CN = 5

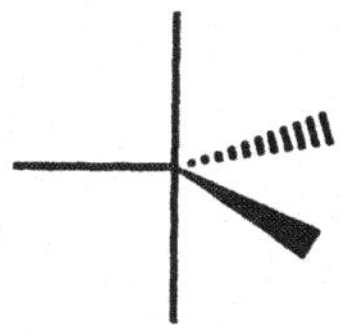

Trigonal Bipyramidal

TB-5†

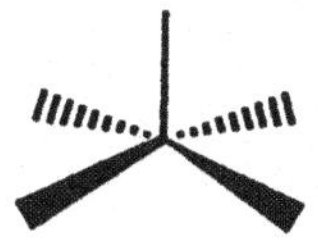

Square Pyramidal

SPY-5

CN = 6

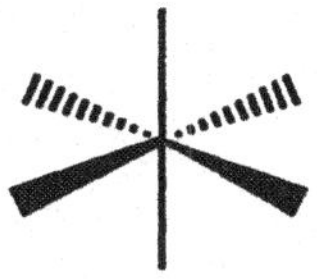

Octahedral

OC-6

Trigonal Prismatic

TP-6

CN = 7

Pentagonal Bipyramidal

PB-7

Mono (square-face-capped)

Trigonal Prismatic

TPS-7

CN = 8

Square Antiprismatic

SA-8

Cubic

CU-8

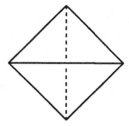 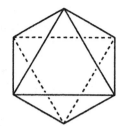

Figure 3.3
Projection graphs for T-4 and OC-6

2–8, they are given in the following list, together with various graphical representations. Drawing programs for chemical formula on computers[‡] give stereo graphs, which emphasize the center-to-ligand connectivity. They are called 'stereo centers' in the particular program used to produced those shown. Such representations can be very useful. Sometimes, however, other ways of representations, stressing the coordination polyhedron are preferable. We give here the stereo centers (Figure 3.2), and projection graphs (Figure 3.3) for the two most important polyhedra, the tetrahedron and the octahedron, where we will often use them instead of the stereo centers. Other types of projection graphs are introduced later for the definition of stereochemical descriptors (Chapter 4, Section 4.3).

Coordination polyhedra for CNs 9–12 have been discussed from a theoretical point of view by Kepert [8].

Coordination symmetry (CS) will be discussed in some detail in Chapter 4. As a first step we want to concentrate on the question of why a particular chemical element in a given molecule has one CG or another.

3.2   Main Group Elements

Main Group elements, i.e. the elements of Groups 1, 2 and 13–18 of the Periodic Table have a strong tendency to form closed-shell molecules (in most cases all the electrons are paired and the compounds are diamagnetic). The CGs of most of such compounds can be predicted in a qualitative way, using surprisingly simple theoretical models. These models give mostly correct answers, even in those relatively rare cases where unpaired electrons occur, i.e. molecules such as NO_2 and ClO_2 and organic free radicals.

3.2.1   Theoretical Models and Predictions

As pointed out earlier, we want to consider mainly 'chemical' theories of chemical bonding, since the experimental chemist, who creates new molecules or tries to

---

[‡]The drawings in this book were made with CSC ChemDraw Plus of Cambridge Scientific Computing.

understand the behavior of molecules in complicated systems, needs a sound qualitative or semi-quantitative theory of molecular structure that puts structures in relation to each other. Such a theory, mainly applicable to Main Group elements, has been developed on the basis of ideas first formulated by Sidgwick and Powell [12] and later brought into a systematic form by Gillespie and Nyholm [13] and others [14]. This theory is known today as *Gillespie–Nyholm* theory or as the *VSEPR* (valence state electron pair repulsion) model. It has been applied to many CGs and to complexes (in an extended form) of transition metals and to structures containing chelate rings and coordination numbers by Kepert [8]. Recently other models, which take into account inter-ligand repulsion and inter-ligand attraction, have been discussed [15].

The VSEPR theory, as its acronym indicates, considers the repulsion between electron pairs, bonding or non-bonding, around a given atom. The physical basis, dictated by the Pauli principle, is the inherent tendency of electrons for pairing in non-degenerate cases, combined with the Coulombic repulsive interaction among electrons. There are three basic rules which apply:

(i)   The coordination geometry is obtained by the minimization of the repulsion between valence state electron pairs.
(ii) The repulsion in a molecule decreases in the following order: $lp/lp > lp/bp > bp/bp$, where lp stands for a non-bonding pair (a lone pair) and bp for a bonding pair.
(iii) The higher the difference in electronegativities between two bonding partners is, the lower is the repulsive power of the corresponding electron pair.

Rule (i) puts two electron pairs in an L-2 geometry, three pairs in TP-3, four pairs in T-4, five pairs in TB-5, six pairs in OC-6, and eight pairs in SA-8. Rules (ii) and (iii) indicate how these geometries may become distorted when not all pairs are equivalent, i.e. in coordination centers carrying non-bonding pairs or in heteroleptic coordination centers. If the existence of a non-bonding pair is stereochemically evident (as it is usually the case), this pair is said to be *stereochemically active*.

Simple as these rules are, they give results that are at least qualitatively correct in most molecules containing only Main Group elements [14] and even in some classes of transition metal compounds [8]. The reader who is interested in more details of these theoretical models is referred to the two books cited, and especially the treatment of Burdett [16]. Here, we give only an overview of the experimental data and their relations to the VSEPR model.

Another theoretical method that has been applied for the simulation of certain aspects of stereochemistry of coordination compounds is molecular mechanics. This method originated in the conformational analysis of organic molecules and it consists essentially in potential energy functions attributed to changes in the internal coordinates of a molecule. The ensemble of such parametrized functions is called the force field of the molecule. For simple organic molecules the application of this method has become routine, and it requires no special input by the user. For coordination compounds, on the other hand, the problems of setting up the force field are still considerable. The subject has been recently reviewed in depth [17].

3.2.2   Experimental Data

CG L-2

This is a relatively rare case with Main Group elements. It occurs in some Be compounds in the gas phase (3.1) and in sp-hybridized C and N centers.

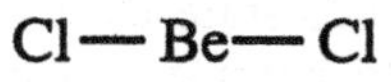

$$Cl—Be—Cl$$

(3.1)

CG A-2

This coordination geometry is widely adopted by non-metallic elements, especially by Group 16 elements (the most prominent molecule is H_2O), but it is hardly known for metallic coordination centers (3.2). The A-2, rather than the L-2 CG of O in H_2O, and in many other related molecules is due to the presence of two non-bonding electron pairs at the oxygen center. For coordination compounds it is important to consider the full stereochemistry of such atomic centers, since they are generally the ligand atoms in coordination compounds. The consequences of the CG of the ligand atom in coordination compounds will be discussed later. A-2 CG is also frequent with sp^2-hybridized N and C centers, which often are ligands in metal complexes.

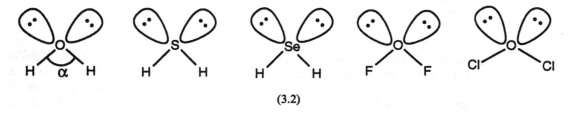

(3.2)

CG TP-3

This geometry occurs in a number of molecules containing a Group 13 element as the central atom, especially with boron compounds (3.3). These molecules are generally strong Lewis acids, owing to the electron pair acceptor property of such 'unsaturated' centers.

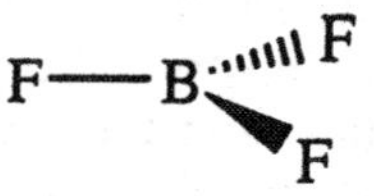

$$F—B\overset{\cdots\text{\tiny III}}{\underset{\blacktriangledown F}{}}F$$

(3.3)

CG TPY-3

In a similar manner to A-2, this coordination geometry is important for elements (Group 15) which are the donor atoms in ligands in coordination compounds. The pyramidal CG (3.4) is due to the presence of one non-bonding electron pair. NH_3 and (almost) all its derivatives show this CG, in addition to the phosphines. Again, it is important to consider the TPY-3 coordination of these centers, as in, e.g. the amines, when they act as ligands in coordination compounds.

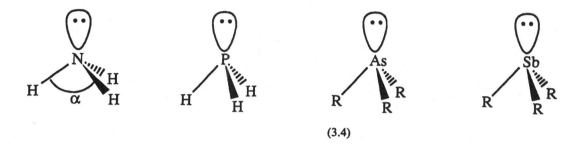

(3.4)

CG T-4

This is the geometry that is adopted by carbon centers in millions of organic molecules. T-4 is for carbon often a highly rigid geometry, where ligands do not exchange places within the lifetime of a chemist. The exchange rate of hydrogen atoms (by tunelling) in methane has been estimated by Hund (quoted in Ref. 18) to be $10^{-20}s^{-1}$. This slow exchange, together with generally slow substitution processes, sets the carbon center apart from other elements where T-4 also occurs. T-4 is attained with elements of Group 15, if the non-bonding pair has become a bonding one, as e.g. when an amine acts as a ligand in a coordination compound or even more simply when it is protonated in an ammonium salt.

The T-4 CG is frequently observed in multiple-bonded complex species (mostly ions) with oxo- and thio-ligands. Examples are $[SO_4]^{2-}$, $[S_2O_3]^{2-}$, $[ClO_4]^-$, $[PO_4]^{3-}$, and a large number of derivatives of these and of similar species.

It also occurs widely in coordination centers of metallic elements (3.5), most notably in complexes of Be^{II} ($Be[acac]_2$), B^{III} (BF_4^-, etc.), Al^{III} ($AlCl_4^-$), Ga^{III}, In^{III}, and Tl^{III} and in Group 14 with Si^{IV}, Ge^{IV}, Sn^{IV}, and Pb^{IV}.

(3.5)

CG SP-4

SP-4 is rare outside the transition elements. It occurs according to the VSEPR model in molecules of the type AL_4E_2, where E denotes a non-bonding pair. A well known example of this is XeF_4 (3.6). There have been many attempts to force atoms, especially carbon, into this CG by putting constraints in the molecule which direct the surroundings of a center into a square. The discussion of these stereochemical questions is outside the scope of this book.

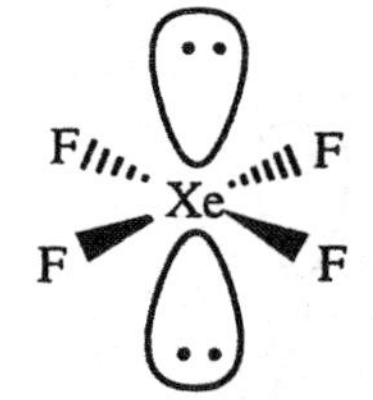

(3.6)

CG TB-5 and SPY-5

Five-coordinate atoms of the Main Group elements are known, especially for the elements of Group 15 in the +V oxidation state. A simple example is PF_5. It is a characteristic of this CG that it is stereochemically non-rigid. The VSEPR model prefers TB-5 CG for PF_5. In penta-coordination, the ligands cannot all be equivalent. They fall into two groups: three equatorial and two axial F-ligands, as shown in Figure 3.4. From this arrangement, the ligands can move to a position in which the molecule appears unchanged, except for a (clockwise) rotation of 90° about an axis perpendicular to the $^1F_e \ldots P$ connecting line. The molecule was not rotated, however, but the ligands were exchanged from equatorial (e) to axial (a) positions and vice versa in the following way:

$$^1F: e \rightarrow a \quad ^2F: e \rightarrow e \quad ^3F: e \rightarrow a$$

$$^4F: a \rightarrow e \quad ^5F: a \rightarrow e$$

This transformation is called a *pseudorotation* [19] (Figure 3.4). Results from computational methods (density functional theory) show that the transition state has SPY-5 CG. It lies only $<10 kJ mol^{-1}$ above the TB-5 configuration (Schafer, O. and Daul, C. personal communication).

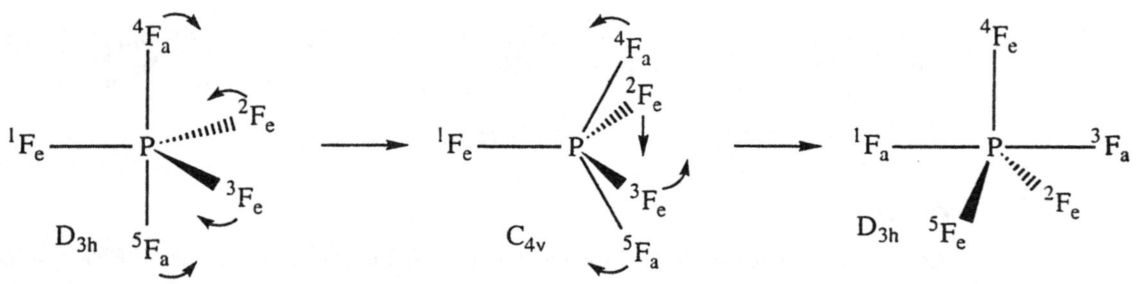

Figure 3.4
TB-5 molecule: PF_5 and the pseudo-rotation

Through a similar and energetically degenerate transformation, also ^{2}F, which is equatorial (e) *before and after* the particular pseudorotation depicted in Figure 3.4, will change to an axial (a) position. Any physical method which is slow on the time-scale of these transformations will give time-averaged values for the observables. This is the case, for example, for ^{19}F NMR spectroscopy, which indicates an apparent equivalence of all five F ligands of P over the temperature range $+60$ to $-197°C$ [18]. This behavior is typical for five-coordinate centers.

CG OC-6

This is by far the most common CG for six-coordinated species and it occurs for non-metallic and metallic centers of the Main Group elements, as e.g. SF_6, PF_6^-, etc. (3.7), and in many complexes of the Group 2, 13, and 14 elements (with the exception of C and Si, which have mainly T-4 coordination). The coordination number and consequently the coordination geometry of such complexes is often simply determined by the ratio of the ionic radii of the ligands and the central ion. Interesting cases of similar stabilities of different coordination geometries arise if the ratio of the ionic radii is close to the critical value, as it is in the case of Al^{3+} as central ion and Cl^- as ligand. The CG of the Al center is T-4 for Al in liquid $AlCl_3$, which contains Al_2Cl_6 molecules, but it is OC-6 in the solid state, where it forms a three-dimensional lattice. Structural aspects of inorganic solids are discussed in a recent textbook by Müller [20].

A case of CN = 6, where OC-6 should not occur according to the VSEPR model,

(3.7)

is $[XeF_6]$, which has a total of seven valence state electron pairs and hence it should form a *distorted* octahedron (Ref. 21, p. 590).

Coordination Numbers > 6

Main Group elements do form species where the CN is larger than six in numerous cases. As far as coordination compounds are concerned, this is mostly the case for alkali and alkaline earth elements with ligands. With these elements the ligands, such as crown ethers or cryptands, predetermine often the CN and the CG of the center. The relation between specific ligands with coordination geometry and symmetry is the subject of Chapter 4, and some examples of such complexes will be discussed there.

3.3   Transition Elements

The 30 transition elements of Groups 3–12 of the Periodic Table (3d, 4d and 5d elements) have an extremely rich and important chemistry, on which a great deal of coordination chemistry is focused. The consequences of the participation of the d-orbitals in the bonding in transition metal compounds are highly variable chemical reactivities, especially also with respect to reduction/oxidation reactions, various coordination numbers and coordination geometries, and highly variable optical, magnetic, and other physical properties. Coordination compounds of these elements are therefore of great significance in many fields of chemistry, including biochemistry, where transition elements play a key role for many important processes. The study of the functions of such elements is a field of great current interest, and the use of some elements that are not normally present in biological systems for diagnostic and therapeutic purposes has become highly important. In many of these studies, fundamental or applied, stereochemical questions are involved.

The repulsion model yields, in general, coordination geometries in agreement with experimental observations for oxidation states corresponding to d^{10} and d^0 configurations. Therefore, coordination units having Sc^{III}, Ti^{IV}, V^V, Cr^{VI}, Mn^{VII}, Cu^I, Zn^{II} or the corresponding elements of the second and third rows as central metals, can be treated in general by the VSEPR model. There are, however, indications that complexes with d^0 metals with strong s-donor ligands may undergo a so-called second-order Jahn–Teller distortion (Ref. 22, p. 95] from VSEPR geometries [23].

The repulsion model can give, in many other cases of transition metal complexes, useful structural information, as was shown in the treatment by Kepert [8]. One of the physical pillars of the VSEPR model is, however, often absent in transition metal compounds, namely the pairing of the electrons. Owing to degeneracy or near degeneracy of d-orbitals, and partial occupation of these levels, unpairing of electrons is a common phenomenon in transition metal chemistry. Some problems concerning stereochemistry cannot be solved without taking into account explicitly the electronic structure of the molecules. Here we are mainly interested in questions such as what coordination number and what coordination geometry occurs in specific cases, for example, why the d^6 ions Fe^{II}, Co^{III}, Ru^{II}, Rh^{III}, Os^{II}, Ir^{III}, and Pt^{IV}, almost exclusively form complexes with OC-6, but SP-4 complexes with the d^8 ions Ni^{II}, Pd^{II}, Rh^I, Pt^{II}, and Au^{III}.

3.3.1. Theoretical Models and Predictions

Ligand field theory is now treated in every textbook of inorganic chemistry. The principal reasons for the widespread occurrence of color among transition metal complexes, and the basis for the magnetic behavior of such compounds, are consequently known by every student in chemistry. *Ligand field stabilization* of various coordination geometries and of different electron configurations is also generally discussed. Ligand field stabilization (this may be CFSE or MOSE, according to Ref. 16) may be an important factor for the CG of a particular complex, but it is generally not the only one. Others, such as the Jahn–Teller effect or even direct steric influences such as the bulkiness of the ligands may also contribute strongly to the CG adopted by a particular metal in a given complex.

A particularly simple ligand field model is the angular overlap model (AOM), which is well suited for the discussion of relative stabilities of different coordination geometries [16]. For example, it does reveal the strong preference of low-spin d^6 configurations for the CG OC-6 and the widespread occurrence of SP-4 for d^8 configurations [24].

In the forthcoming discussion, it will be assumed that the reader is familiar with the basic concepts of ligand field theory, and it is recommended to go through the arguments given by the AOM in order to understand the preferences of a given central metal for certain coordination geometries, as given in Table 3.1. The information contained in this table should be treated as a rough guide to reality, giving only the most frequent coordination numbers and coordination geometries.

Table 3.1 Coordination polyhedra for transition elements in coordination species. Only the most common coordination geometries are given

		d^0			
Sc^{III} OC-6	Ti^{IV} T-4 OC-6	V^V T-4 OC-6	Cr^{VI} T-4	Mn^{VII} T-4	
Y^{III} SA-8	Zr^{IV} T-4 OC-6	Nb^V CN 4 to 9 variable	Mo^{VI} **T-4** OC-6	Tc^{VII} T-4	Ru^{VIII} T-4
La^{III} CN 4 to 11	Hf^{IV} PBP-7 + variable	Ta^V variable	W^{VI} T-4 OC-6	Re^{VII} T-4 + variable	Os^{VIII} T-4
		d^1			
Ti^{III} OC-6	V^{IV} T-4 OC-6 SPY-5 (VO^{2+})	Cr^V T-4 OC-6	Mn^{VI} T-4		
	Nb^{IV} OC-6	Mo^V CN = 8	Tc^{VI} OC-6	Ru^{VII} T-4	
	Ta^{IV}	W^V CN = 8	Re^{VI} OC-6	Os^{VII} T-4	
		d^2			
Ti^{II} OC-6	V^{III} T-4 **OC-6**	Cr^{IV} T-4 OC-6	Mn^V T-4	Fe^{VI} T-4	
			Tc^V OC-6	Ru^{VI} OC-6	
			Re^{VI} OC-6	Os^{VI} T-4	

continued overleaf

Table 3.1 *Continued*

d^3

Cr^{III} OC-6	Mn^{IV} OC-6	(Fe^V)
Mo^{III} OC-6	Tc^{IV} OC-6	
W^{III} OC-6	Re^{IV} OC-6	

d^4

Cr^{II} OC-6 (dist.)	Mn^{III} OC-6	(Fe^{IV}) OC-6
	Tc^{IV} OC-6	Ru^{IV} OC-6
	Re^{IV} OC-6 TP-6	Os^{IV} OC-6

d^5

Cr^{I} OC-6	Mn^{II} T-4 **OC-6**	Fe^{III} T-4 **OC-6**	Co^{IV} OC-6
		Ru^{III} OC-6	Rh^{IV} OC-6
	Re^{II} OC-6	Os^{III} OC-6	Ir^{IV} OC-6

d^6

Cr^{0} OC-6	Mn^{I} OC-6	Fe^{II} T-4 **OC-6**	Co^{III} OC-6	Ni^{IV}
Mo^{0} OC-6		Ru^{II} OC-6	Rh^{III} OC-6	Pd^{IV} OC-6
W^{0} OC-6	Re^{I} TB-5 OC-6	Os^{II} OC-6	Ir^{III} OC-6	Pt^{IV} OC-6

d^7

Fe^{I} OC-6	Co^{II} T-4 OC-6	Ni^{III} TB-5 OC-6 (dist.)

Rare earth elements and actinides occur in coordination compounds with a large variety of coordination numbers and coordination geometries. The coordination numbers often exceed six. Semi-empirical methods for the prediction of CNs and CGs are virtually non-existent, and *ab initio* methods often inconclusive. In

Table 3.1 *Continued*

	d^8		
Co^I	Ni^{II}	Cu^{III}	
TB-5	SP-4	SP-4	
	OC-6	OC-6	
Rh^I	Pd^{II}	Ag^{III}	
SP-4	SP-4	SP-4	
Ir^I	Pt^{II}	Au^{III}	
SP-4	SP-4	SP-4	

	d^9	
Ni^I	Cu^{II}	
T-4	T-4	
	SP-4	
	TB-5	
	OC-6 (dist.)	
Pd^I	Ag^{II}	
SP-4	SP-4	
	Au^{II}	
	SP-4	

	d^{10}	
Ni^0	Cu^I	Zn^{II}
T-4	T-4	T-4
		OC-6
	Ag^I	Cd^{II}
	L-2	OC-6
	T-4	
	Au^I	Hg^{II}
	L-2	L-2
		T-4

addition, most metal to ligand bonds in these classes of metals are highly labile. Isomerism is therefore not common for complexes of rare earth and actinide elements. Consequently, stereochemistry is to a large extent metric stereochemistry, mainly determined by an experimental approach trough X-ray crystallography.

3.4   References

1. Leigh, G. J. (1993), *Chem. Br.*, **29**, 574.
2. Kauffman, G. B. (1993), *Chem. Br.*, **29**, 867–868.
3. Muetterties, E. L. and Guggenberger, L. J. (1974), *J. Am. Chem. Soc.*, **96**, 1748–1756.
4. Eliel, E. L. and Wilen, S. H., *Stereochemistry of Organic Compounds*, Wiley–Interscience, New York, 1994.
5. Berman, M. (1971), *J. Franklin Inst.*, **291**, 229–260.
6. Britton, D. and Dunitz, J. D. (1973), *Acta Crystallogr,. Sect. A*, **29**, 362–371.

7. Federico, P. J. (1975), *Geometriae Dedicata*, **3**, 469–481.
8. Kepert, D. L., *Inorganic Stereochemistry, Inorganic Chemistry Concepts*, Vol. 6, Springer, Berlin, 1982.
9. King, R. B. (1991), *J. Math. Chem.*, **7**, 51–68.
10. Leigh, G. J., *Nomenclature of Inorganic Chemistry*, Blackwell, Oxford, 1990.
11. Block, B. P., Powell, W. H. and Fernelius, W. C., *Nomenclature of Inorganic Chemistry. – Recommendations*, ACS Professional Reference Book, Blackwell, Washington, DC, 1990.
12. Sidgwick, N. V. and Powell, H. M. (1940), *Proc. R. Soc. (London), Ser. A*, **176**, 153–180.
13. Gillespie, R. J. and Nyholm, R. S. (1957), *Q. Rev. Chem. Soc.*, **11**, 339–380.
14. Gillespie, R. J. and Hargittai, I., *The VSEPR Model of Molecular Geometry*, Allyn and Bacon, Boston, 1991.
15. Rodger, A. and Johnson, B. F. G. (1992), *Inorg. Chim. Acta*, **191**, 109–113.
16. Burdett, J. K., *Molecular Shapes. Theoretical Models of Inorganic Stereochemistry*, Wiley, New York, 1980.
17. Hay, B. P. (1993), *Coord. Chem. Rev.*, **126**, 177-236.
18. Muetterties, E. L. (1970), *Acc. Chem. Res.*, **3**, 266-273.
19. Berry, R. S. (1960), *J. Chem. Phys.*, **32**, 933-938.
20. Müller, U., *Inorganic Structural Chemistry, Inorganic Chemistry: A Textbook Series*, Wiley, Chichester, 1994.
21. Cotton, F. A. and Wilkinson, G., *Advanced Inorganic Chemistry*, 5th. edn, Wiley, New York, 1988.
22. Albright, T. A., Burdett, J. K. and Whangbo, M.-H., *Orbital Interactions in Chemistry*, Wiley, New York, 1985, p. 95.
23. Kang, S. K., Tang, H. and Albright, T. A. (1993), *J. Am. Chem. Soc.*, **115**, 1971–1981.
24. Purcell, K. F. and Kotz, J. C., *Inorganic Chemistry*, Saunders, Philadelphia, 1977.

4 General Concepts in Topographic Stereochemistry of Coordination Compounds

4.1 Symmetry

In this chapter we want to consider the arrangements of the atoms in coordination compounds in some detail, i.e. we are interested not only in the coordination geometry of a given center, but also in the number and kind of different isomers, and their 'exact' symmetry. For that purpose, we need to define a number of general concepts and we propose a classification scheme for ligands that will be useful for further discussions.

Chemical compounds consisting of discrete molecular units (where these units can be charged, i.e. the 'molecular' units can be ions) and which can be crystallized, show two kinds of symmetry: the symmetry of the crystal and that of the molecular unit itself. Not all chemical compounds have this property, e.g. simple lattice compounds such as the alkali metal halides do not consist of molecular units. The symmetry of the crystal and that of the molecular units forming it can have certain relations to each other. For example, an enantiomerically pure chiral molecule can never crystallize in a centrosymmetric space group. Crystal and molecular symmetry are, however, two clearly distinct properties. Here we are mainly concerned with molecular symmetry, and we will only occasionally refer to the symmetry of the crystal of a coordination compound.

The molecular symmetry is different from the coordination geometry, in that it takes into account the *nature* of the ligands around a coordination center and their exact positions. A T-4 coordination center e.g. has tetrahedral symmetry if all four ligands of the center are *identical*. Even this condition does not suffice for exact T_d symmetry. If the complex is for example $[M(H_2O)_4]$,[†] it cannot have exact T_d symmetry (Figure 4.1), since the ligand H_2O does not comply with all the symmetry elements of T_d. The H_2O ligands desymmetrize the complex to a maximum symmetry of C_{2v}. On the other hand, $[MF_4]$ can have exact T_d symmetry. For most considerations however, it is an excellent approximation to assume T_d symmetry for $[M(H_2O)_4]$.

Symmetry can be treated mathematically by group theory. The application of group theoretical methods in chemistry for the discussion of various molecular properties, such as derivation of electronic energy diagrams and analysis of

---

[†]In order to simplify the writing of formulae, we do not give the charge of a complex unless it is a concrete species. for example we write $[Fe(H_2O)_6]^{2+}$, but $[M(H_2O)_6]$ or $[Ml_6]$.

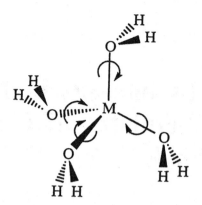

Figure 4.1
[M(H$_2$O)$_4$] complex. T-4 at the M-center, TPY-3 at the O-center

vibrational modes, is treated in a number of excellent text books [1]. It was pioneered by physicists shortly after the advent of quantum mechanics [2]. Coordination chemists are generally familiar nowadays with group theoretical concepts and with group theoretical nomenclature. Group theoretical arguments will only be occasionally used here, and a thorough knowledge of this method is therefore not necessary. It is necessary, however, that the reader is familiar with the basic concepts of molecular symmetry. Therefore, a short summary of terms is given here (Figure 4.2).

- A *symmetry operation* is a transformation, carried out or imagined on a physical model of the molecule, which puts the model in a congruent position in space.
- A *symmetry element* is a geometric object (point, axis, plane), which is invariant during the symmetry operation.
- There are basically two different types of symmetry operations: *proper* and *improper rotations*. Proper rotations can be carried out on a physical model. Improper rotations can only be imagined.
- The symmetry elements corresponding to proper rotation (Figure 4.2) are the *n*-fold rotational axes C_n, where rotation through an angle $2\pi/n$ is a symmetry operation.
- The symmetry elements corresponding to improper rotations (Figure 4.2) are the invariant geometric objects obtained when a molecule is rotated through an angle of $2\pi/n$, followed by an imagined mirror operation through a plane perpendicular to the axis. These elements (the axes of improper rotation) are designated as S_n. S_1 is a mirror plane (symbol σ) and S_2 is a center of inversion (symbol i). S_n for $n > 2$ are axes in the molecule.

The symmetry operations of a molecule (including the identity operation $C_1 \equiv E$) form a group in the mathematical sense.

For molecular symmetries the use of the Schoenfliess nomenclature is generally preferred over the Herrmann–Mauguin nomenclature used in crystallography. For the convenience of the reader, a list of the Schoenfliess designations and the

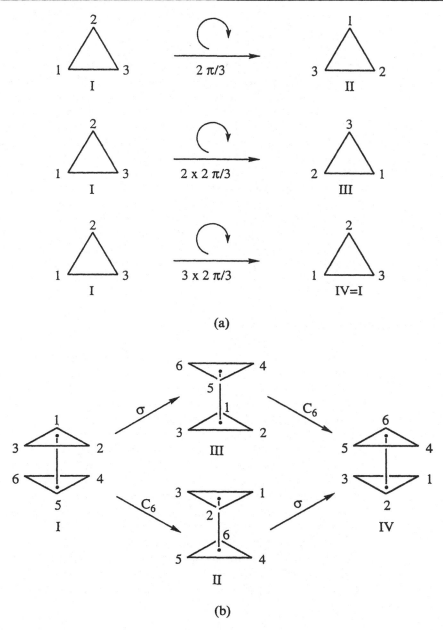

(a)

(b)

Figure 4.2
Representations of (a) proper and (b) improper rotations (Adapted from [1]). (a) *Proper rotation (C₃)*: configurations II and III are equivalent to I because without the labels (which are not real, but only our mental constructions) they are indistinguishable from I, although with the labels they are distinguishable. However, IV is indistinguishable from I not only without labels but also with labels. Hence it is not merely equivalent, it is *identical*. (b) *Improper rotation (S₆)*: II and III are equivalent to each other but neither is equivalent to I; neither σ nor C₆ is by itself a symmetry operation, but the combination of both, in either order, which we call S_6, is a symmetry operation since it produces IV, which is equivalent to I

corresponding symmetry elements for frequently encountered molecular symmetry groups is given in Appendix I.

Group theoretical tables contain the characters of the irreducible representations and they feature transformation properties of functions, etc. For such group theoretical applications the reader is referred to tables of this kind [3].

Until now, we have discussed the symmetry of a whole molecule, which we will always call *molecular symmetry*. For certain purposes it is useful and even necessary to consider *local symmetries* within a segment of a molecular model.[†] If we consider a segment that is occupied by exactly one atom, we speak of the *site symmetry* at the atom [5]. In molecules with D_n or higher symmetries, only one atom in the molecule can have the full molecular symmetry group as its site symmetry. All other atoms will have a lower site symmetry. In C_n molecular symmetry, all atoms lying on the C_n axis have the full molecular symmetry as their site symmetry, all other atoms have C_1 site symmetry.

As an example, consider an OC-6 complex $M(a_1)_6$ having a molecular symmetry of O_h (a_1 is the symbol of a monoatomic ligand). Only the central metal M has O_h site symmetry, whereas the a_1 Ligands have C_{4v} site symmetry.[‡] Figure 4.3 shows site symmetries for various atoms in OC-6 and SP-4 complexes.

4.2   The Classification of Ligands

The coordination center as defined in Chapter 3 (p. 20) is one point in space and therefore, without the ligands, it is an entity without a three-dimensional structure. The *ligands* are the *structure-forming elements* in a coordination unit. For the discussion of the stereochemistry of coordination compounds, it is therefore useful to introduce a classification scheme of ligands. The classification scheme introduced here corresponds to the general usage by coordination chemists, giving some additional, more precise, information when needed.

It is neither necessary nor useful to divide the ligands into *mutually exclusive* classes. A given ligand may therefore very well appear in two or more of the classes defined below. As we are not considering metal clusters, the ligands discussed here will never contain a metallic element as the *ligating* atom. Although we exclude most organometallic compounds from our discussion, carbon as ligating atom will be included and discussed.

The number of ligands known today is exceedingly large and it grows every day (a significant part of all *ca* 10^7 chemical species registered by Chemical Abstracts Service (CA) are at least potential ligands). It is therefore not really possible to

[†]Important relationships between the local symmetry of a segment and the molecular symmetry are: (i) the restriction that no segment may contain a symmetry element that does not belong to the molecular symmetry, (ii) all segments of a chiral molecule are chiral, and (iii) segments of an achiral molecule may be chiral or achiral. For further discussion the reader is referred to Ref. 4.
[‡]This matters if a nuclear property, which is sensitive to the site symmetry, is measured. For example the hyperfine coupling tensor in a paramagnetic complex for the M-nucleus in the above-mentioned complex will be isotropic, i.e. a scalar quantity (due to the cubic site symmetry), whereas that of a_1 nucleus will have axial symmetry, i.e. two principal values, with the principal axis coinciding with the M–a_1 direction. As in this example, the site symmetry group of the metal center in a coordination compound is often that of the molecule itself, whereas the site symmetry group of the ligand atoms is a genuine subgroup of the former.

(a)

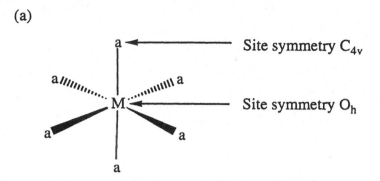

(b)

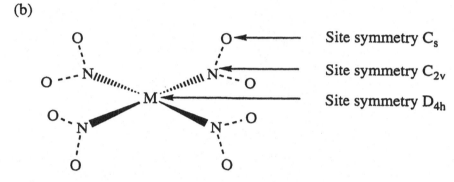

Figure 4.3
(a) Site symmetries in a molecule of overall symmetry O_h. (b) Site symmetries in a molecule of overall symmetry D_{4h} (SP-4), with four AB_2 ligands (NO_2^-)

present a comprehensive list of ligands. A small selection is given in Appendix III, in order to familiarize the reader with the various types of ligands available today, and also for the sake of introducing abbreviations used in this book.

4.2.1 Ligating Atoms

We define as the ligating atom the one directly connected to the coordination center. The ligating atoms in a coordination species therefore constitute the first coordination sphere. Ligating atoms in coordination compounds are atoms of the nonmetallic elements in the Periodic Table, generally in a well defined oxidation state. The latter statement, however, needs to be taken *cum grano salis*, since several cases (e.g. NO as a ligand) are known where the oxidation number of the ligating atom is by no means unambiguously definable [6].

The ligating atoms are given in Table 4.1, where roman numerals give the oxidation state.

4.2.2 Number of Atoms in a Ligand

A ligand can possess any number of atomic centers. For stereochemical purposes it is convenient to distinguish three different levels of complexity:

Table 4.1 The most important ligating atoms in coordination chemistry. The oxidation states are given in parentheses

Group 17:	F(−I), Cl(−I), Br(−I), I(−I), At(−I)
Group 16:	O(−II, −I, O), S(−II to +IV), Se(−II to +IV), Te(−II, . . .)
Group 15:	N(−III to +III), P(−III to +III), As(−III to +III)
Group 14:	C(−IV to +II)

(a) Monoatomic Ligands

This group of ligands is not very large and, consequently, it can be enumerated exhaustively (rare cases in parentheses):

$$F^-, Cl^-, Br^-, I^-, O^{2-}, S^{2-}, Te^{2-}, (N^{3-}, P^{3-})$$

If it is necessary to specify a ligand of this group in a formula, we will use a small roman letter with subscript 1, such as a_1, b_1, etc., or in a general case l_1.

The coordination species occurring in cryolite, AlF_6^{3-}, is an example of a metal complex with only monoatomic ligands. It may be written in a general way, for the purpose of discussing its stereochemistry, as $[M(a_1)_6]$.

(b) Monocentric and Polycentric Ligands

Monocentric ligands are those containing only atoms directly connected to the ligating atom. A large number of commonly occurring ligands constitute this group. Only a selection is given below:

$$H_2O, OH^-, O_2, NH_3, N_2, PF_3, NO, NO_2^-, CN^-, \text{etc.}$$

If it is necessary to specify a ligand of this group in a formula explicitly, we will use a small roman letter with subscript m, such as a_m, b_m, etc., or in general l_m. An example is $[Co(NH_3)_6]^{3+}$, which would be written in a general way as $[M(a_m)_6]$. Note that OC-6 $[M(a_1)_6]$ can have exact O_h symmetry, whereas OC-6 $[M(a_m)_6]$ often cannot have exact O_h symmetry (except e.g. in the case where a_m is a linearly coordinated ligand such as CN^-), because certain of the symmetry elements required by O_h will not be present in the ligands. In many cases however, these deviations from an idealized symmetry can be neglected.

Ligands are often more complex, such as polyatomic molecules with complicated connectivity. Such ligands are, e.g.

$$R_2O, RO^-, NH_2R, NHR_2, NR_3$$

If they have only one ligating atom, (see monodentate ligands below), we designate them, if necessary, as a_p, b_p, or in general as l_p.

4.2.3 Ligating Properties of Ligands

(a) Monodentate Ligands

Ligands that can only be connected with one ligating atom or those which *are* connected by one ligating atom only to a coordination center (even if there are other ligating atoms in the ligand) are either called monodentate ligands in the former case, or they bind as monodentate ligands in the latter case. Ligands possess often one and only one ligating atom. This is the case, by definition, for all l_1, and for most l_m ligands, although O_2 as a l_m ligand can act as a monodentate or as a bidentate ligand. In the latter it binds 'side on' to a coordination center.

We will designate monodentate ligands of an unspecified nature (l_1, l_m, or l_p) with small roman letters a, b, c,..., or in general by l.

(b) Chelate Ligands

Chelate ligands, or *chelates* in short, are by definition ligands where the ligating atoms and the coordination center forms at least one closed loop, when the connectivity of the atoms is considered. Chelating ligands must therefore have at least two ligating atoms. The total number of ligand atoms that can bind simultaneously to one metal center is the *denticity* of the ligand. A closed loop containing the metallic coordination center is called the *chelate ring*. The smallest chelate ring has three atoms. For the purposes of stereochemical discussions, chelate ligands will be designated with one capital letter per ligating atom, with a symbol of connectivity (^) signifying 'connected via one or several bonds'. Sometimes, if a connection via a long chain of atoms is present, ^^ will be used.

L will be the symbol of an unspecified ligating atom, and A, B, for specific donors (e.g. N, O, P). A class of chelating ligands that has attracted much attention during the past two decades are those in which at least one σ-donor atom is either sp^2 or sp^3 carbon. These are generally called *cyclometalating* ligands, and the coordination units derived from them are consequently *cyclometalated* complexes.

A bidentate ligand (Figure 4.4) with two equivalent ligating atoms will be symbolized (A^A), with two different ones (A^B), and with general donor atoms (L^L). In certain cases it will be useful to distinguish between such ligands which form planar and non-planar chelate rings, respectively. If this is the case, we designate the planar ring as (A=A), and the non-planar as (A≈A). Examples are 2,2′-bipyridine for (A=A) and 1,2-diaminoethane, in coordination chemistry called *ethylenediamine* (en) for (A≈A).

Multidentate chelating ligands can have many different topologies and it is neither possible, nor really practical, to introduce general symbolic designations for all these cases. In any concrete case, symbols will be used which are self-explanatory, using similar elements as above. For example (A^A^A) can be a symbol for either terpyridyl (c) (Figure 4.4) or for dien (d). The designation (A=A=A), however would clearly mean a ligand such as *terpy* and not *dien*, which would be written as (A≈A≈A). In each concrete case, ligands will be defined by their structural formula and in general an abbreviation will be introduced, since systematic names of many ligand molecules are hopelessly cumbersome. Recently, ligands that are especially designed for large coordination numbers have been described [7].

(c) Macrocyclic and Cage Ligands

An important class of ligands has a constitution in which the atoms in the ligand itself form a closed loop, which contains at least two, but in general more than

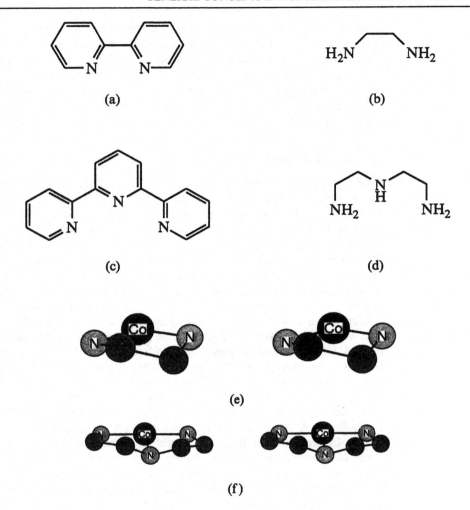

Figure 4.4
Examples of chelates: (a) 2,2′-bipyridine, bpy (A=A); (b) 1,2-diaminoethane, ethylenediamine, en (A≈A); (c) 2,2′:6,2″-terpyridine, terpy (A=A=A); (d) diethylenetriamine, dien, (A≈A≈A); (e) and (f) stereo pairs showing the nonplanarity of the chelate rings in the [Co(en)] and the [Co(dien)] fragments, respectively (see Section 5.3.2.)

two, ligating atoms. These ligands form the class of the *macrocyclic ligands*. Examples are the naturally occurring heme ligand (4.1), crown ethers (4.2), cryptands (4.3), etc. No general designation is attempted for these ligands, except that we will abbreviate a macrocyclic ligand without any further specification as *ane* or *[n]ane*, if we want to specify the ring size.

A special class of macrocyclic ligands are those which have several interconnecting loops with ligating atoms in the loops. These ligands are called *cage ligands* (4.4).

If the cage is not closed, the ligands are often called *half cage* (or *octopus*, see Chapter 5, Section 5.7) ligands (4.5).

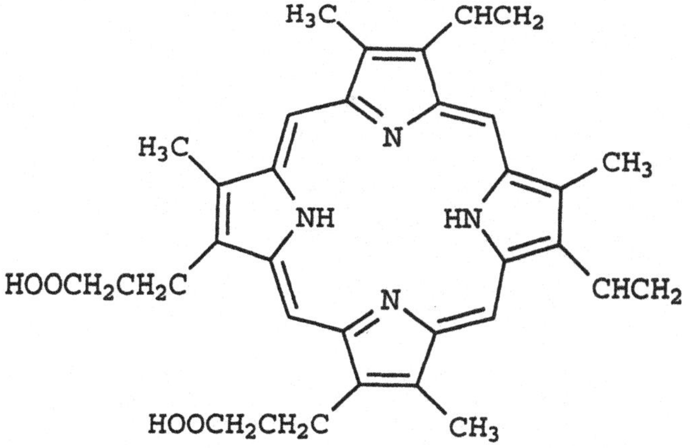

(4.1)

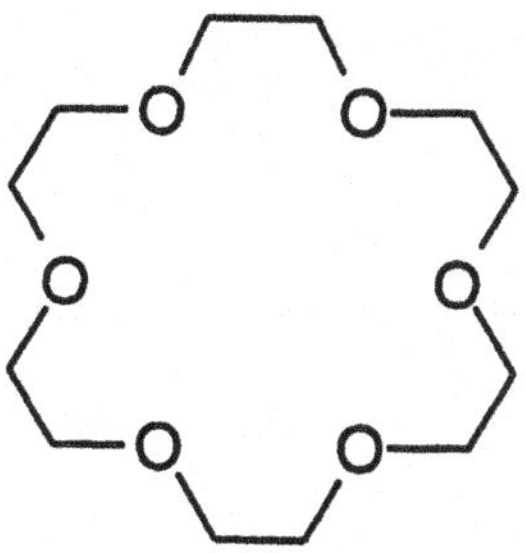

(4.2)

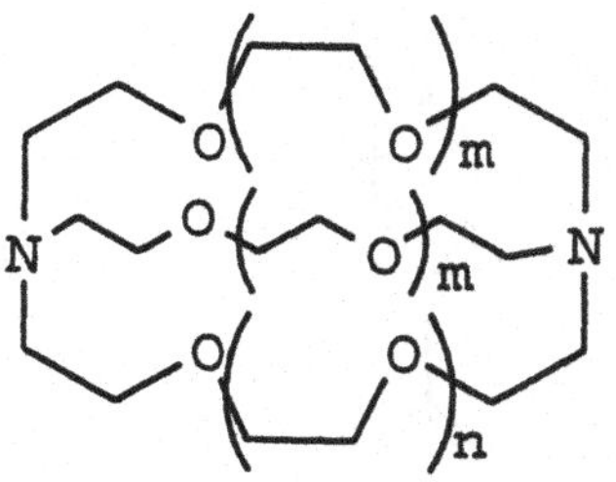

(4.3)

(d) Bridging Ligands, Polynuclear Complexes

If a ligand is bound simultaneously to more than one coordination center, the ligand is a bridging ligand. It can be, in principle, a ligand of any of the above-mentioned classes. Following the official nomenclature, the bridging ligand has a prefix μ- in the formula of the complex. The complexes that have more than one metallic

(4.4)

(4.5)

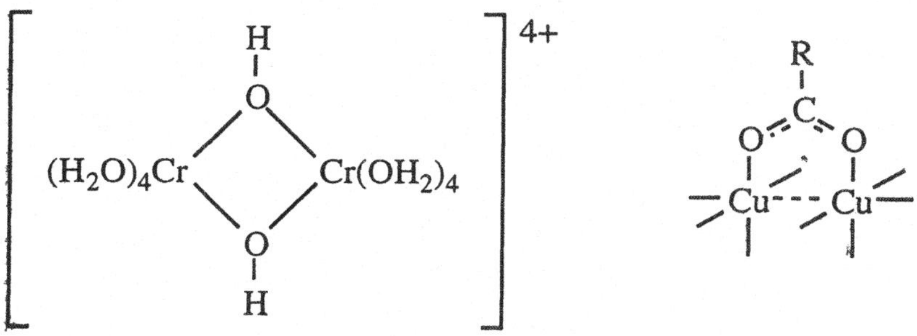

$$\left[(H_2O)_4Cr \underset{\underset{H}{O}}{\overset{\overset{H}{O}}{<}} Cr(OH_2)_4 \right]^{4+}$$

(a)

(b)

(4.6)

coordination center are called polynuclear complexes (di-, tri-, etc.). Examples are given in (4.6).

(e) Ambidentate Ligands

Ambidentate ligands are molecules (or ions), with (at least) two different ligating atoms or ligating sites that do not simultaneously bind to a metallic center. They will be designated as (ab). The coordination species M(ab) is not the same as M(ba), since in the former $\underline{a}$b is the ligating atom and in the latter a$\underline{b}$. This can result in a special kind of isomerism (see below).

Examples of ambidentate ligands are NO_2^- (ligating atoms either N or O) and NCS^- (N or S) (4.7).

$$\begin{cases} M-N\overset{O}{\underset{O^-}{<}} & \text{Nitro} \\ M-ONO & \text{Nitrito} \end{cases} \qquad \begin{cases} M-SCN & S\text{-Thiocyanato} \\ M-NCS & N\text{-Thiocyanato} \end{cases} \qquad \begin{cases} M-CN & \text{Cyano} \\ M-NC & \text{Isocyano} \end{cases}$$

(4.7)

(f) Alterdentate Ligands

Alterdentate Ligands are different from ambidentate ligands insofar as the two ligating sites are equivalent. An alterdentate ligand can consequently be designated as (aa). The two species (I) M(aa) and (II) (aa)M are indistinguishable, but the transition between (I) and (II) is an observable process. Examples of (aa) ligands are simple molecules, such as O_2 and N_3^-, and chelating species such as ninhydrin and alloxane (4.8).

Terpy, as another example of a potentially alterdentate ligand, can act either as a terdentate chelate or as an alterdentate ligand (4.9). Both modes of coordination are known.

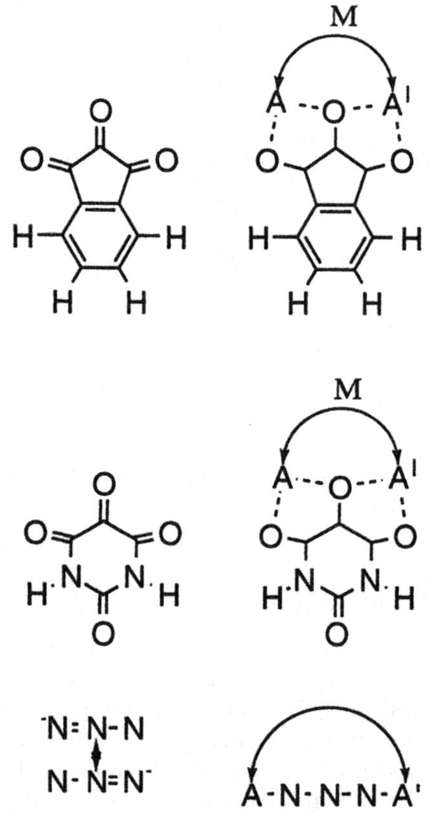

(4.8)

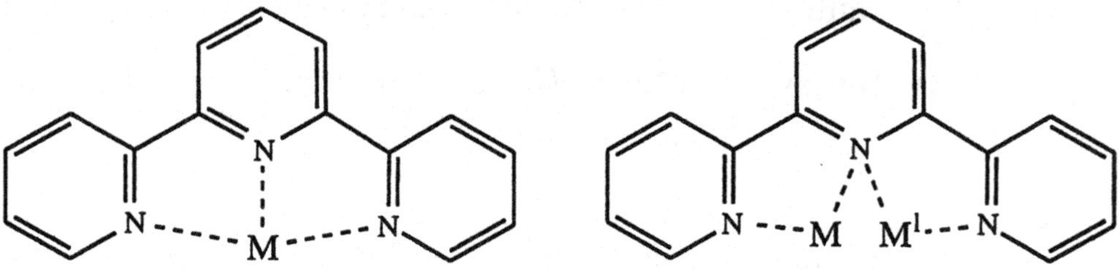

(4.9)

(g) Chiral and Achiral Ligands

Ligands from any of the above mentioned class, with the exception of l_1 (which cannot be chiral), can be either achiral or chiral species when they are detached from the metallic coordination center. *Unless indicated otherwise*, we shall always tacitly assume that a ligand is achiral. In the case that it is chiral, we will indicate it by an appropriate chirality descriptor.

4.3 Isomerism†

4.3.1 General Considerations

In molecular sciences, *isomers* of a given compound are generally defined as substances having the same stoichiometric composition and the same 'molecular weight,' but which can be distinguished by some chemical or physical methods. With molecular weight, we mean here the mass of the unit under consideration. This definition excludes polymorphism of solids to be considered as isomerism, and also molecules with different isotopic composition, which are called *isotopomers*.

With the assumption introduced earlier that a molecular unit under consideration (where the time scale of the observation has to be specified) has assigned to it a well defined connectivity of the atomic centers, all isomers can be divided into two classes: *structural isomers* or *constitutional isomers* and *stereoisomers* or *configurational isomers*, respectively. Constitutional isomers differ in their connectivity, i.e. some atomic centers have different numbers and/or different kinds of ligands (where the centers are any atoms in the molecular unit under consideration), whereas stereoisomers have the same structures but they differ in some other way in the arrangement of the atoms in space.

Definition 4.1

Constitutional isomers are distinguished by a different connectivity of the atoms in the molecular species.

Definition 4.2

Stereoisomers are molecules of identical constitution, but differing in the arrangements of the atomic nuclei in space.

Constitutional Isomers

Constitutional isomers play an important role in coordination chemistry in a somewhat different way than in organic chemistry. In organic chemistry, constitutional isomerism arises often through a different branching of a carbon–carbon bonded skeleton or/and through different positions of functional groups that are bound to such a skeleton.

In coordination chemistry several cases of constitutional isomers have been distinguished by Werner (see Ref. 8 and references cited therein):

---

†An excellent, concise account of isomerism in coordination chemistry is given in the series *Comprehensive Coordination Chemistry* by Harrowfield and Wild [8] and an earlier one treating similar problems by Buckingham [9].

(a) Ionization, hydrate, coordination, and ligand isomerism.

- *Ionization isomers* are compounds where a given charged ligand is either coordinated to the central metal or present as counter ion in the lattice. Examples are numerous Co^{III} complexes studied by Werner and others:

$$[CoCl_2(en)_2]NO_2 \text{ and } [CoCl(NO_2)(en)_2]Cl \text{ or } [CoCl(NCS)(en)_2)]NCS$$
$$\text{and } [Co(NCS)_2(en)_2]Cl$$

- *Hydrate isomers* occur because water is a common ligand in the first coordination sphere, but it is also easily bound in the crystal through hydrogen bonding forces. The best known examples are Cr^{III} complexes:

$$[CrCl_2(H_2O)_4]Cl \cdot 2H_2O \text{ and } [CrCl(H_2O)_5]Cl_2 \cdot H_2O$$

- *Coordination isomers* occur if at least two metals are present, which can accommodate at least two different kinds of ligands in their first coordination sphere in different ways. A simple example is:

$$[Cr(NH_3)_6][Co(CN)_6] \text{ and } [Co(NH_3)_6][Cr(CN)_6]$$

Although the *compounds* designated as ionization, hydrate, and coordination isomers are clearly isomers, the *coordination units* from which they are built, do not have the same molecular weights and they are therefore not isomeric in the sense of our definition.

- An isomerism which is fairly frequent is *ligand isomerism*. In this case the coordination units have the same stoichiometric compositions and the same molecular weights. Connectivities within the ligands are different, however. As an example the complexes given in (4.10) [10] are ligand isomers.

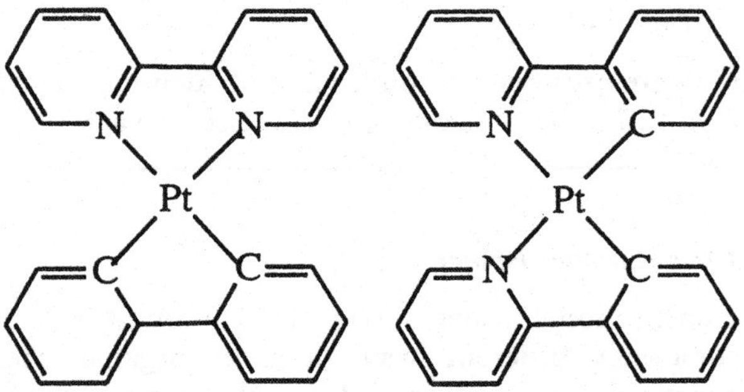

(4.10)

- Another example of ligand isomerism is the three isomeric dinuclear complexes given in (4.11).

(b) Linkage isomers. Linkage isomers occur with ambidentate ligands (ab) (it could also occur with (A^B) if the chelating ligand is bound only as a monodentate ligand).

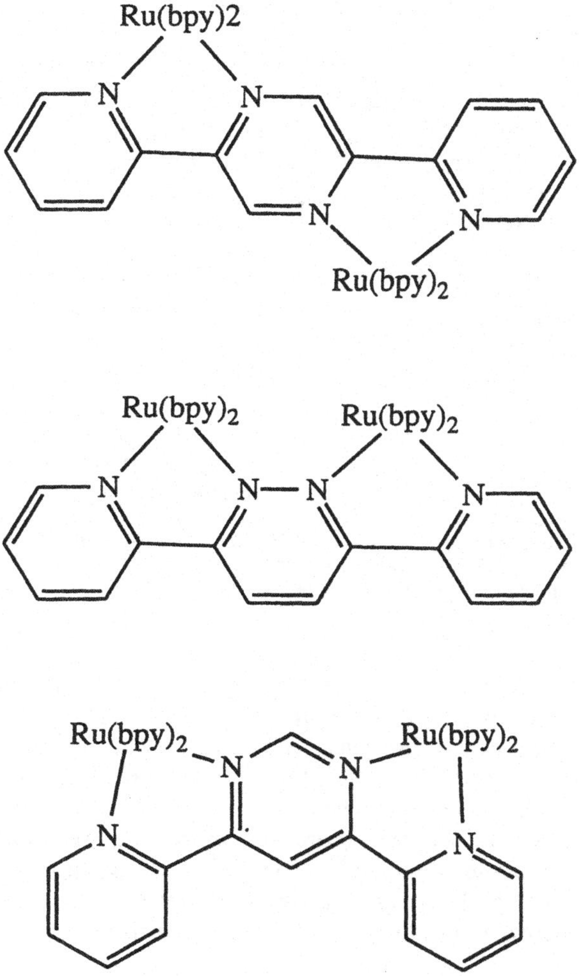

(4.11)

Numerous cases of linkage isomerism with various ambidentate ligands have been described [11]. Examples include the following ligands:

$$NO_2^-, \ SCN^-, \ CN^-, \ DMSO, \ \text{substituted pyridines, etc.}$$

Such isomers yield often important information regarding specificity of binding of a metal to one or another type of ligand atom. Some examples are given in Table 4.2.

Stereoisomers

Stereoisomers can be sharply divided into two subsets, namely into *enantiomers* and *diastereomers* (or *diastereoisomers*).

Table 4.2 Examples of linkage isomers with NO_2^-, NCS^-, and CN^- ligands. for details, see Refs 11 and 12[a]

M-ONO and M-NO$_2$	M-SCN and M-NCS	M-NC and M-CN
$[Co(NH_3)_5NO_2]^{2+}$	$[Pd(As(C_6H_5)_3)_2(NCS)_2]$	$[Co(CN)_5CN]^{3-}$
$[Co(NH_3)_2(py)_2(NO_2)_2]^+$	$[Pd(bpy)(NCS)_2]$	$[Cr(H_2O)_5CN]^{2+}$
$[Co(en)_2(NO_2)_2]^+$	$[Cd(CNS)_4]^{2-}$	cis-a-$[Co(trien)(CN)_2]^+$
$[Rh(NH_3)_5NO_2]^{2+}$	$[Mn(CO)_5SCN]$	
$[Ir(NH_3)_5NO_2]^{2+}$	$[Rh(NH_3)_5NCS]^{2+}$	
$[Pt(NH_3)_5NO_2]^{3+}$	$[Ir(NH_3)_5NCS]^{2+}$	
$[Co(CN)_5NO_2]^{3-}$	$[Cr(H_2O)_5NCS]^{2+}$	
$[Ni(Me_2en)_2(ONO)_2]$	$[Pd(Et_4dien)NCS]^+$	
$[Ni(EtenEt)_2(ONO)_2]$	$[Pd(4, 7\text{-diphenylphen})(SCN)_2]$	
	$[Cu(tripyam)(NCS)_2]$	
	$[(C_5H_5)Fe(CO)_2NCS]$	
	$[(C_5H_5)Mo(CO)_3NCS]$	
	$[Pd(P(OCH_3)_3)_2(NCS)_2]$	

[a]Abbreviations: Me$_2$en = N,N-dimethylethylenediamine; EtenEt = N,N'-diethylethylenediamine; bpy = 2,2'-bipyridine; 4,7-diphenylphen = 4,7-diphenyl-1, 10-phenanthroline; tripyam = tri-(2-pyridyl)amine (bidentate).

(a) Enantiomers.   Enantiomers are sometimes called *optical isomers,* owing to their ability to rotate the plane of linearly polarized light in opposite directions. Optical isomerism is, however, one of the several misnomers [13] used in many textbooks of chemistry in connection with stereochemical subjects. *Catoptromer and catoptric* as adjective have been proposed [14] as alternative expressions. However, their usage has become more and more rare. In addition, these terms have not been applied in organic chemistry. They do not appear in the recent book by Eliel and Wilen [15]. As far as possible, chemists should try to use convergent language. We therefore discourage the use of catoptromer and catoptric.

Definition 4.3

An enantiomer is one of a pair of molecules which are mutually congruent with the realized form of their mirror image. An enantiomer is not congruent with its own mirror image.

From a fundamental point of view, the two isomers obtained by space inversion are not true enantiomers, since they are not strictly degenerate because of parity violation of the weak interaction forces. The exact opposite, which is the meaning of the term enantiomer, is to be found in the antiworld, i.e. a space-inverted molecule composed of antimatter [16]. For chemistry this does not matter, since the non-degeneracy is extremely small and molecules composed of antimatter are beyond the wildest dreams of chemists in the matter world. It can matter, however, for theories of the origin of homochirality in biological systems (Ref. 15, p. 209*ff*). Such theories are undoubtedly highly interesting and they pose a real intellectual challenge. They are, and may even remain, unprovable hypotheses given the fact that evolution has been probably a *unique event*.

Enantiomerism is a dichotomic phenomenon. If a molecule shows enantiomerism, there is always exactly one pair of enantiomers. The concept is closely related to that of chirality.

Definition 4.4

A molecule or any geometric object is said to be chiral[†] if the object and its realized mirror image are not congruent [20].

A chiral molecule can therefore exist always in two different forms, one being the realized form of the mirror image of the other. These two forms are the enantiomers. Later chirality descriptors will be discussed for various cases. Since the two enantiomeric forms of a chiral molecule always rotate the plane of linearly polarized light of a given wavelength in opposite directions, they can be designated as the (+) and the (−) forms. Enantiomers will therefore be labeled either by chirality descriptors if the absolute configuration is to be addressed, or by (+) or (−), or by both. If (+) and (−) are used in a specific way, the reference wavelength has to be specified.

The adjective chiral does not imply that molecules are present in one enantiomeric form only. If a substance is composed of chiral molecules, the two enantiomers may be present in any composition given by $0 \leqslant |ee| \leqslant 1$ (see p. 54), where $ee = 0$ corresponds to the racemate and $|ee| = 1$ to an enantiomerically pure compound.

(b) Diastereomers. We have stated before that stereoisomers are either diastereomers or enantiomers. Therefore, we can define now diastereomerism by a negating phrase:

Definition 4.5

Stereoisomers that are not enantiomers are diastereomers (diastereoisomers).

The scheme (4.12) that we can apply is the following (connections with dashed lines express decisions in the affirmative): ① is the connectivity in the molecules the same? ② is the mirror image congruent to the molecule?

Diastereomers behave generally differently with respect to most chemical and physical properties. Diastereomers are therefore clearly considered to be different chemical compounds. Diastereomers are often divided again into two groups: *configurational isomers* and *geometrical isomers*. Beside the fact that geometrical isomer is really another misnomer (because geometrical can mean anything that relates to geometry and thus to all stereochemistry), this distinction is not

---

[†]A synonym for chiral is dissymmetric, which was used by Pasteur as designation for objects that differ 'only as an image in a mirror differs from the object which produces it' [17–19]. We prefer chiral, because it defines the targeted property in a positive way, whereas dissymmetric literally translates as 'lack or absence of symmetry'. Chiral objects may, however, have certain elements of symmetry.

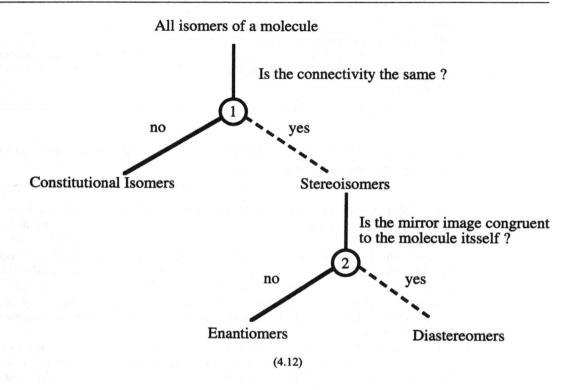

(4.12)

unambiguous since it is based on a difference in energy barrier between two or more isomeric forms. Although geometrical isomer is a term still often used by coordination chemists, it should be replaced by more specific and meaningful designations. There are many different kinds of diastereomers in coordination compounds which will be discussed in subsequent chapters.

One rather special type of diastereomers, occurring with some transition metal complexes, is *polytopal* isomerism. Isomers that occur because two polyhedra with the same coordination number have similar energies have been called *polytopal* [21] *isomers* or *allogons* [8,22]. In practice, it is not always easy to decide whether polytopal isomers exist, especially in solution where interconversion of two coordination geometries may be very fast, especially in the case of TB-5/SPY-5 where polytopal isomerism could be expected for energetic reasons. However, in solids, polytopal isomerism has been shown to exist. An example is $[Ni(CN)_5]^{3-}$, where TB-5 and SPY-5 was shown to exist in the solid [23]. Both complexes have five nickel to carbon bonds, and consequently the same connectivity. They are therefore not constitutional isomers, but stereoisomers. Being achiral, they are clearly diastereomers.

(c) Enantiomers again. As opposed to diastereomers, the two enantiomers of a pair have not always been considered to be different chemical compounds, since they behave identically in many respects. Three different cases can be distinguished:

• Enantiomers behave identically in an achiral environment (where a molecule of the same kind is not considered to be part of the environment).

- Enantiomers behave differently in a non-racemic chiral environment.
- Enantiomers in a racemic chiral environment give opposite signs for certain observables. Each of the two enantiomers of a pair can interact in a different way with the constituents of the racemic chiral environment, where the interaction obeys a pairwise identical relation.

A chiral environment is a system that contains chiral constituents. It can be either racemic if it contains 'exactly' the same number[†] of molecules of opposite chirality (of each kind, if there are several types of chiral molecules), or non-racemic, if one enantiomer is predominant. The constituents can be either molecules or, e.g., circularly polarized photons.

Some examples will show what is meant by these three cases:

(i) Suppose one or the other (pure) enantiomer of a chiral compound either is present as a pure substance, or it is dissolved in a solvent consisting of non-chiral molecules (as is the case for most solvents). A molecule will than be in a surrounding that is either locally achiral (the solvent molecules), or that has the same chirality as the molecule itself (another molecule of the compound). Observables, which do not depend on chiral probes, such as thermodynamic properties, interactions with non-polarized electromagnetic fields (polarized electromagnetic radiation has to be in a certain wavelength range, for the appearance of measurable effects on its polarization), or chemical reactivities with achiral substrates, will be, under these circumstances, identical for the two enantiomers of a chiral compound.

(ii) The two enantiomers of a chiral compound will behave differently in many of their properties in a non-racemic chiral environment. With respect to non-racemic environments, enantiomers are therefore to be considered as *chemically distinct compounds*. The most important non-racemic chiral environment is formed by the biosphere. In general, enantiomers therefore behave differently, sometimes even dramatically differently in biological systems. Drug-licensing agencies have become aware of this fact, and they consider today enantiomers to be chemically different compounds, and mixtures of enantiomers are not considered to be pure substances [24]. The synthesis of pure enantiomers (EPC = enantiomerically pure compounds[‡]) has therefore become a major task, especially in the pharmaceutical industry [26].

The differences in properties of the two enantiomers of a chiral compound in a non-racemic chiral environment provided by living systems range from different rates for biodegradation to different sensory qualities to the tragic case of mutagenicity, now known to exist for only one enantiomer of thalidomide [15, pp.201, 204; 27].

---

†We assume that we have always samples containing a large number of molecules, so that statistical deviations become negligible.

‡'Enantiomerically pure = enantiopure' is a limiting hypothesis, which has to be used with the same caution as the concept of a pure chemical compound in general. There have been various propositions for concepts expressing quantitatively the enantiomeric purity of a compound [25], [15, p 161]. If necessary, we will use the enantiomeric excess, defined below, as a measure for enantiomeric purity. A terminology for an enantiomerically 'pure' compound, that should be avoided, is homochiral. This term should only be used in comparing the chiralities of two or several chiral centers in a molecule.

(iii) The best known case of an interaction of one enantiomer with a racemic environment is the interaction of an enantiomer with linearly polarized light. Linearly polarized light is a racemate of right- and left-circularly polarized light. One enantiomer will rotate the plane of the polarization of light (at a given wavelength) in one direction and the other enantiomer by the same amount in the other direction. This is the well known *optical activity* of non-racemic chiral compounds. A pure enantiomer or a non-racemic mixture of the two enantiomers of a chiral compound is therefore called an *optically active substance*. Optical activity should not be used synonymously with chirality. Only a macroscopic sample having unequal concentrations of the (+) and the (−) enantiomers of a substance consisting of chiral molecules is optically active (there might be an accidental absence of rotational power at a given wavelength). A racemate does not rotate the plane of polarized light, and it should therefore not be called an optically active substance.

The specific interaction of one enantiomer with one constituent of the racemic mixture may be, e.g., the chemical reactivity. The pairwise relationship means that the following interactions are identical: $(+)_{\text{solute}}/(+)_{\text{environment}} \equiv (-)_{\text{solute}}/(-)_{\text{environment}}$, $(+)_{\text{solute}}/(-)_{\text{environment}} \equiv (-)_{\text{solute}}/(+)_{\text{environment}}$.

The purity of an enantiomer with respect to the 'contamination' with its partner of opposite absolute configuration is generally given by the *enantiomeric excess* (ee), which is a dimensionless quantity defined by

$$(\text{ee}) = \frac{x_{(+)} - x_{(-)}}{x_{(+)} + x_{(-)}}$$

or this value multiplied by 100, if ee is to be given in % [$x_{(+)}$ and $x_{(-)}$ are the mole fractions of the two enantiomers, respectively]. In general, ee is defined to be $1 \geqslant \text{ee} \geqslant 0$, i.e. it is always referred to the major enantiomer. Sometimes, however, it is practical to give ee values over the entire range of compositions. In this case (ee) values vary between −1 [pure (−)-form] to 0 (racemate) to +1 [pure (+)-form].

Enantiomers should be considered to be different chemical compounds, with some special relationship to each other. This special relationship implies a symmetry in diagrams that plot the value of measurements of observables of chiral compounds in non-chiral environments as functions of the (ee). Such diagrams must be symmetric with respect to the racemate with (ee) = 0. Examples of such diagrams are melting points of enantiomeric mixtures and solubilities in non-chiral solvents. This symmetry requirement can be fulfilled in three different ways:

(i) Racemic mixtures. If the melting point and solubility diagrams for enantiomeric mixtures have a minimum for the melting point and a maximum for the solubility, (Figure 4.5), the racemate is called a racemic mixture, with the eutectic point at (ee) = 0.

The thermodynamic reason for this kind of behavior is a higher Gibbs free energy of the racemate as compared to the EPCs. On the molecular level, the molecules of one enantiomer interact more favorably than the other when forming a solid with the same enantiomer.

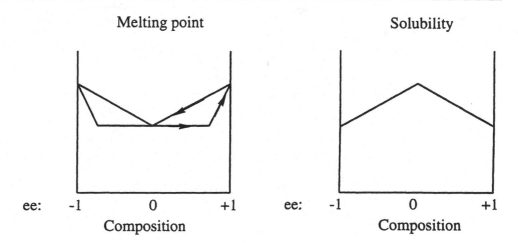

Figure 4.5
Phase and solubility diagrams for racemic mixtures. For details see Ref. 28

Racemic mixtures crystallize as *macroscopic conglomerates* [29] of pure crystals of the (+)- and of pure crystals of the (−)-forms, as e.g. the famous sodium ammonium tartrate studied by Pasteur. Many coordination compounds, crystallizing as conglomerates, have been prepared and investigated by Bernal [30,31]. Edith Humphrey's compound, prepared in 1900 [32,33] represents an example (p. 2).

(ii) Racemic compounds, racemic modifications. If the melting point of the system decreases when either of the EPC is added to the racemate, the latter is to be considered as a true compound. This compound is often called the racemic modification of the enantiomers [28]. The racemic compound and the EPCs generally form two eutectic mixtures at some intermediate composition (Figure 4.6), and the racemic compound will have minimum solubility.

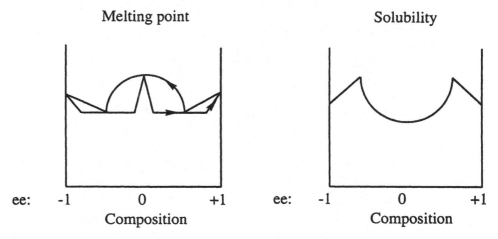

Figure 4.6
Phase and solubility diagrams for a racemic compound

On the molecular level, this means that enantiomers are interacting more favorably with their partners of opposite absolute configuration than with themselves. This case is more frequent than that of enantiomeric mixtures. The crystals obtained for a racemic compound contain an equal number of molecules of both enantiomers in the unit cell.

In both cases, racemic mixtures and racemic modifications, ordered crystals will generally be obtained by crystallization from solutions containing racemates. Racemic mixtures will crystallize as *conglomerates of crystals*, which must belong to a *non-centrosymmetric (or generally enantiomorphic)* space group. Crystallization of a chiral substance in an enantiomorphic space group[†] does not, however, indicate necessarily that the molecules in one crystal are homochiral [34]. If the number of molecules in the unit cell is an even multiple of the number of symmetry positions in that space group, it is possible that the crystal contains a racemic compound, i.e. an equal number of the two enantiomers occupying non-symmetry-equivalent sites in the crystal [35]. Even if the number of molecules in the unit cell of an enantiomorphic space group is an odd integer (>1) of the number of symmetry positions, there is no guarantee that all molecules are homochiral. It still is possible that the crystal contains an unequal number of molecules of opposite chiralities, as observed in several Rh, Ir, and Cu complexes [36–38]. In this case one crystal will contain neither a racemic compound nor an EPC, but one enantiomer enriched according to its presence in the unit cell of the crystal, i.e. $0 < (ee) < 1$. A racemate of a substance behaving in the aforementioned way crystallizes again as a conglomerate, but not one consisting of crystals of EPCs. Table 4.3 gives the relationships between the requirements for the space group in the solid state and the type of compound.

Crystals of chiral molecules having non-enantiomorphic space groups must, however, contain an equal number of both enantiomers of the chiral molecules and they are therefore a clear indication of the presence of a racemic modification of the substance.

(iii) Racemic solid solutions. If the two enantiomers of a chiral compound form ideal or nearly ideal solid solutions, their melting point and solubility diagrams will

Table 4.3 Relationship between properties of the space group of solids composed of chiral molecules.

Type of system	Requirement for space group in the crystalline state
EPC	Enantiomorphic
Racemic mixture	Two enantiomorphic forms of the same space group (conglomerate) either all molecules in one enantiomorphic form are homochiral, or the ratio of both enantiomers is $\neq 1$
Racemic compound	Non-enantiomorphic space group, or enantiomorphic space group, with the number of molecules in the unit cell being an even multiple of the number of symmetry positions.

[†]Space groups lacking an inversion center and mirror planes, as well as higher order improper rotation axes are called enantiomorphic. Crystals having such space groups can occur in two macroscopically enantiomorphic forms.

be horizontal straight lines. In this case no preference exists in the interaction for one or the other form and all mixtures of the two enantiomers have the same Gibbs free energy. In this case a solid solution will form upon crystallization of the racemate. The solids can be true crystals in the sense that they are built up in a regular way with the unit cell as building blocks, but they will be disordered with respect to the distribution of the two enantiomers.

4.3.2 Chirality and Prochirality in Coordination Compounds

Although the general definition of chirality is straightforward and unambiguous,[†] many concepts related to chirality are not always simple. It was a triumph of the tetrahedral model of the carbon atom by van't Hoff and Le Bel in 1874 that it provided a simple theoretical basis for the discovery of enantiomeric chemical compounds by Pasteur in 1848. Since 1874 the 'asymmetric' carbon atom, i.e., by definition, a carbon atom having four different ligands, played a central role in the discussion of chiral organic compounds. This connection between a property of the whole molecule (chirality) and a local property of a center in a molecule (asymmetric atom) has for a long time given rise to some non-trivial conceptual problems.

A rather obvious fact was realized early on: molecular chirality is by no means restricted to molecules having asymmetric atoms in the sense of the definition given above. Asymmetry, i.e. the absence of any symmetry element except the identity operation, is *not* a requirement for chirality. The requirement in terms of molecular symmetry can be easily stated: any molecule lacking an improper symmetry axis S_n is chiral, although proper rotational axes of any rank may be present. This can be easily seen on a macroscopic model: an n-bladed propeller ($n > 1$) has a n-fold symmetry axis C_n and it is a chiral structure. Of special practical importance is the case of S_1 and S_2: molecules having either a mirror plane ($S_1 = \sigma$) or a center of inversion ($S_2 = i$) cannot be chiral. Werner was perfectly aware of the difference between the chirality caused by an asymmetric carbon atom and that induced by the chelate ligands around an octahedral metal center, because he mentioned this in his first publication about chiral metal complexes. Although an OC-6 complex $M(A^\frown A)_3$ can have (and generally will have) D_3 symmetry, and with that two threefold axes and three twofold symmetry axes, Werner speaks of such a metal complex as an 'asymmetric atom' [40].

A more subtle difficulty with the concept of an asymmetric atom arises if a carbon atom $[C(a)(b)\{C_R(x,y,z)\}\{C_S(x,y,z)\}]$ is considered, where R and S are the usual chirality descriptors[‡] for an 'asymmetric carbon atom.'

Although the central C atom of such a molecule is asymmetric, following the usual definition that an asymmetric T-4 atom has four different ligands, it lies on a plane of symmetry, containing the a and b ligands and bisecting the $C_S C C_R$ angle, and it is therefore not a center of chirality. For this case the term *pseudo-asymmetric* carbon atom was coined, but a really satisfactorily clarification for this and other problems with so called 'elements of chirality' was reached only 1984.

[†]This statement is only pertinent for geometrical objects at rest. If time-dependent phenomena are considered, the definition of chirality has to be refined [39].
[‡]The chirality descriptors will be discussed in Section 4.4.

Mislow and Siegel [41] showed that the traditional linkage between stereoisomerism and local chirality leads to conceptional confusion. The term *stereogenic center* is currently used to designate a T-4 coordinated atom that was formerly called an *asymmetric atom* [42].

The concept of a *stereogenic center* is more general than that of the corresponding 'asymmetric atom' used for a long time in organic chemistry. An atomic center in a molecule is called stereogenic if an interchange of two ligands leads to a stereoisomer. If the stereoisomer obtained is the other enantiomer of a pair, the atom is a chiral center.

(4.13)

The cobalt centers in (4.13) are stereogenic [43], but not chiral. Exchange of Cl^- and NH_3 leads to the *cis* and *trans* diastereomers, respectively. In order to distinguish these two subsets of stereogenic centers, the concept of *chirotopicity* was introduced [41]. A stereogenic center is *chirotopic* if its site symmetry group contains no element of improper rotation, otherwise it is *achirotopic*. The site symmetries of the two cobalt centers in (4.13) are D_{4h} and C_{2v}, respectively. The cobalt center is therefore achirotopic. Any true asymmetric T-4 center has C_1 site symmetry, and it is therefore chirotopic. Stereogenic centers may, or may not be centers of chirality, but all chiral centers are stereogenic.

Elements of Chirality

The concept of the *chiral center* (a point in space) that leads to enantiomerism (a single configurationally stable chiral center in a molecule, is a sufficient, but not a necessary condition for enantiomerism of the molecule), is generally brought into relation with T-4 or TPY-3 centers [15, p. 1194]. It can easily be generalized to centers of higher CN. All chiral coordination units with only monodentate ligands can be designated as chiral centers. On the other hand, complexes with one or several chelate ligands, proper symmetry axes of any rank might appear.

In organic chemistry, the concept of chiral elements was later generalized to include *chiral axes* and *chiral planes*, which can be applied to enantiomeric molecules devoid of chiral centers. Organic molecules [15, pp. 1120–1121] having chiral axes and planes are given in Figure 4.7.

These three elements of chirality in 3D space, which have themselves the dimensions 0, 1, and 2, are treated in the classical publication by Cahn, Ingold and

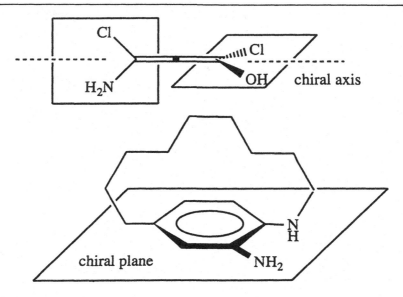

Figure 4.7
Examples of an axis and a plane of chirality

Prelog [44], and in subsequent publications [45]. Structures comprising *chiral axes* or *chiral planes* can alternatively be viewed as helical. In coordination chemistry, most chiral structures are *devoid of chiral centers*. It is rather the rule than the exception that a helical chirality occurs and this point is therefore emphasized in most considerations of chiral coordination compounds.

Organic chemistry can be regarded as a multicenter coordination chemistry with (very often tetrahedrally coordinated) carbon atoms being the coordination centers. Carbon is one of the very few T-4 coordination centers that is substitutionally inert, giving rise to isolable isomers, which are distinguished only in the arrangements of the ligands around the coordination center, as it is the case for an 'asymmetric carbon atom.' For practical purposes, consideration of elements of chirality is therefore very important for organic chemistry, where the segmentation of molecules into subunits is the usual way of looking at them. Coordination compounds are also very often partitioned into segments (this partitioning is a purely formal process, and does not correspond to real chemical fragmentation of the molecule). The obvious partitioning is into central metal and ligands in the first place. In coordination compounds chirality can have its origin in different segments of the molecule: it can be caused by (Figure 4.8): (i) the different nature of simple ligands, as in the stereogenic (asymmetric) carbon atom, and in [Mabcdef]; (ii) by the formation of a helical structure when chelate ligands 'wrap' around the metal; (iii) by chiral conformations within a chelate ring; (iv) by ligands that become chiral upon coordination, because an improper symmetry is broken; (v) by coordination of chiral ligands; and (vi) if two ligands coordinated to a metallic center break mutually elements of improper rotation present in each individual ligand.

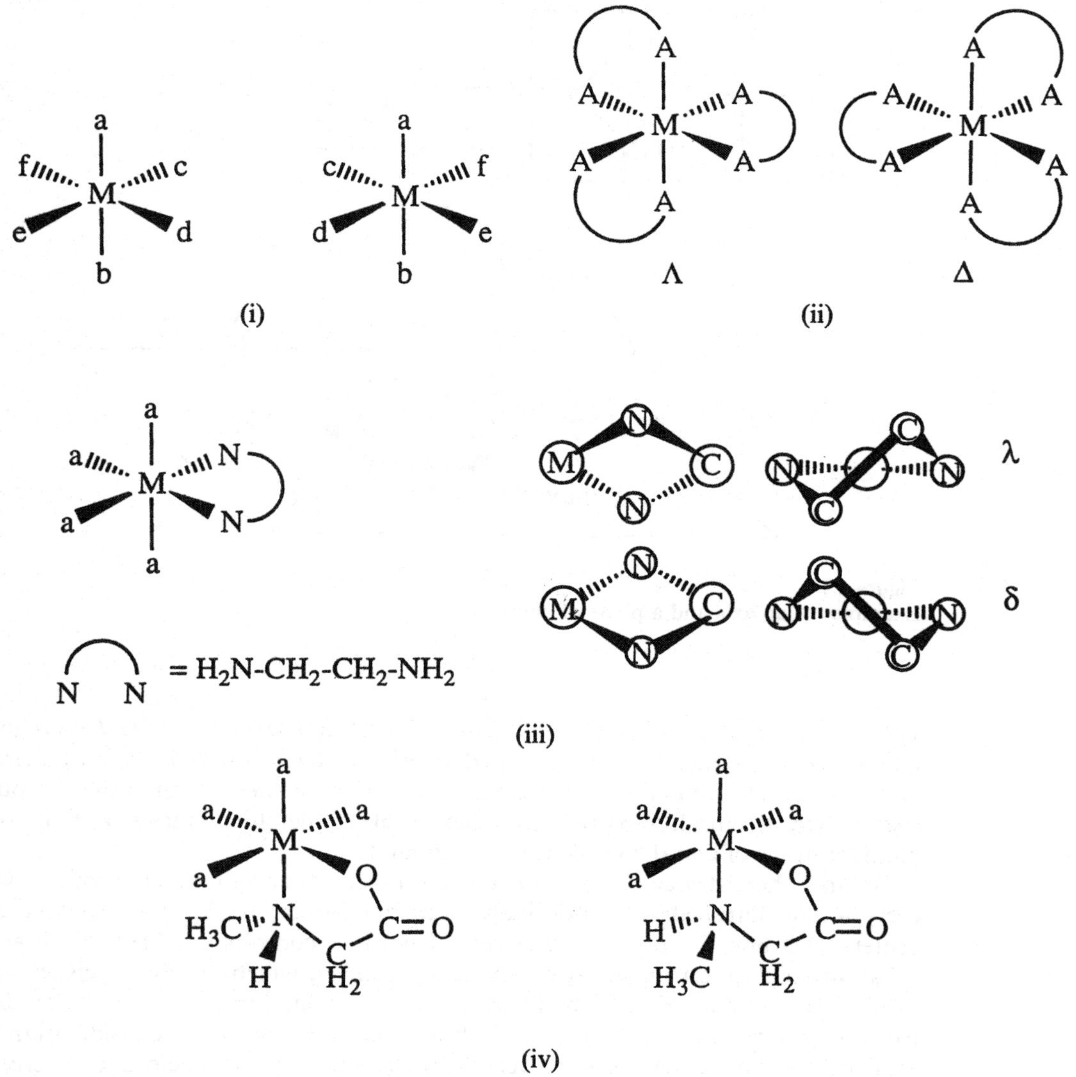

Figure 4.8
Six different origins of chirality in coordination compounds: (i) [Mabcdef]; (ii) [M(A^A)₃]; (iii) [M(en)a₄];
(iv) [M(A(R₁;R₂;R₃)^B)a₄] : [M(sarcosinato)a₄]; (v) [M(alaninato)a₄]; (vi) [Pt(isobutylenediamine)(*meso*-stilbenediamine)]²⁺; from Ref. 46

Prochirality

Consideration of molecular structure is one important aspect of chemistry, but even
more important is the consideration of transformation of chemical structures. It is
therefore useful to look at molecules within the perspective of their transformation
in chemical reactions. A molecule that is not chiral can, in certain cases, become
chiral in chemical reactions, or it can form two or more diastereomers. On this

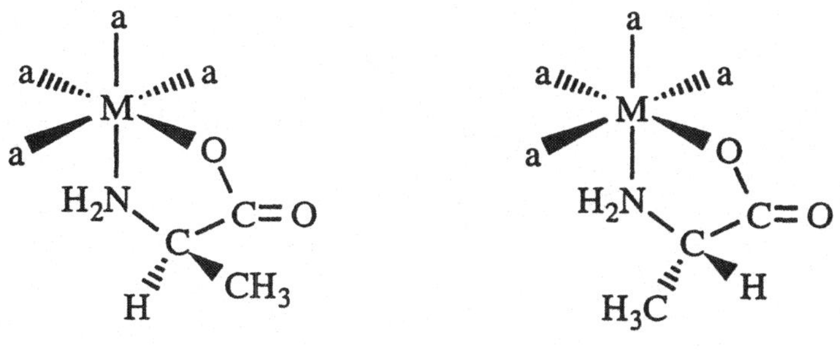

(v)

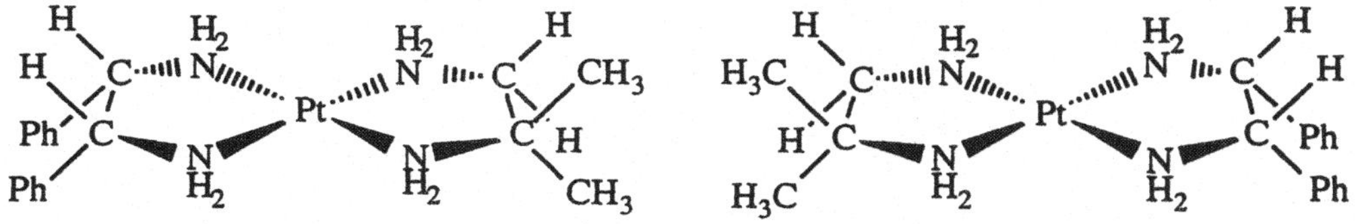

(vi)

Figure 4.8 *Continued*

basis the concepts of *prochirality* [47] and later that of *prostereoisomerism* [15, p. 465], were created. Two examples (Figures 4.9 and 4.10), taken from organic chemistry, show clearly what is meant by these two concepts.

Prochirality and prostereogenicity are properties of a molecule as a whole. It is also possible to identify elements of prochirality and prostereogenicity (center, axis, and plane) [15, p. 466]. As discussed by Mislow [41], the use of *elements of prochirality* gives rise to similar problems as elements of chirality, e.g. the *asymmetric carbon* atom.

In general we will, in accordance with most discussions of stereochemical properties of coordination compounds, treat coordination compounds from a more holistic point of view. Local symmetry concepts will, however, be used in many concrete cases, especially in connection with ligand conformation and chiral ligands and in the application of the concept of *prochirality* to coordination compounds, which was recently discussed by Mestroni *et al.* [48].

4.4 Stereochemical Nomenclature

Molecular sciences deal with highly diversified and often complicated structures. A well defined language to describe its objects, i.e. the molecules, is therefore a necessary base of understanding chemistry. The first international conference in chemistry, the Geneva Conference in 1892, dealt mainly with questions of nomenclature in organic chemistry. There the basis was laid for the language of organic chemistry as it has developed until

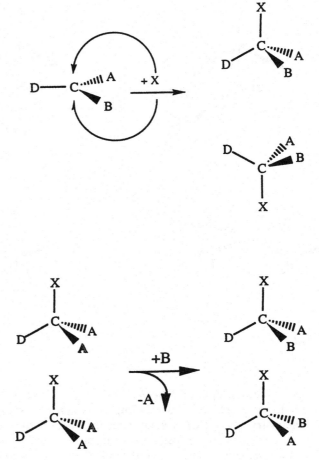

Figure 4.9
Two cases of prochiral molecules: top, TP-3; bottom, T-4

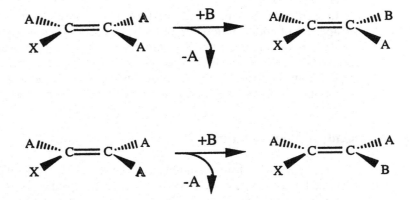

Figure 4.10
The prostereogenic molecule AXC=CA$_2$

today. Inorganic chemistry may in the past have generally dealt with less complicated molecules, but it is in other respects more complex than organic chemistry. The Commission on the Nomenclature of Inorganic Chemistry of the International Union of Pure and Applied Chemistry (IUPAC) began to study the problem of nomenclature in inorganic chemistry in 1921. Since 1940, when its first rules appeared, IUPAC has steadily developed the system of nomenclature of inorganic compounds. The most recent publication is the 1990 *Recommendations on Nomenclature of Inorganic Chemistry* [49], (the so-called *Red Book*, Part I). It treats coordination compounds and their stereochemistry in some detail (Chapters 1–10). A book on *Inorganic Chemical Nomenclature* was also published by the American Chemical Society in 1990 [50]. It treats stereochemistry very similarly to the *Red Book*, but it is not in all respects in complete accordance with the latter. For stereochemical nomenclature it contains a useful appendix (Chart A.2, pp. 178– 186), which gives full stereochemical descriptors in a condensed form. The basis for the use of stereochemical descriptors was laid by a system proposed in a publication of Brown *et al.* [51], which was extended to higher coordination numbers by the same authors [52, 53]. Systematic chemical nomenclature can be of great value for understanding certain concepts, but it can also become very cumbersome for real molecules. In this chapter, we want to concentrate on certain, especially stereochemical, aspects of nomenclature of coordination chemistry, but we will not apply official IUPAC rules to their full extent for more complicated cases. In addition, we will introduce concepts such as the oriented line reference system for chiral structures, which are not 'officially' endorsed by IUPAC. We do this for practical purposes and not with the intention of disobeying IUPAC or of trying to prejudice future decisions of the official IUPAC Commission.

Stereochemical descriptors: The stereochemical descriptors of mononuclear coordination compounds with only monodentate ligands [M(a)(b)(c)...] comprise essentially three elements: (i) the polyhedral symbol (called symmetry site term in Ref. 50) for the coordination geometry, (ii) a coded symbol (either a number or a location term), describing the diastereomeric arrangement of the ligands (the number is called *configuration number* in Ref. 50 and *configuration index* in the *Red Book*, and (iii) a chirality symbol assigning the absolute configuration.

For chelate, macrocycle, cage, etc., complexes, more detailed rules are needed for a systematic nomenclature. In the *Red Book* of 1990, such rules are given for certain types of chelates (ligands with two ligating atoms are in general usage *bidentate* ligands, in the *Red Book* they are designated *didentate* ligands), but not for macrocyclic compounds or for more complicated connectivities. The three elements mentioned above are important for understanding the stereochemistry of coordination units and we will therefore discuss (ii) and (iii) here.

Polyhedral symbols: The significance of the polyhedral symbols has already been given in Chapter 3 (p. 22/23). Here we want to stress once more that these symbols generally do not give the site symmetry of the coordination center, but rather an idealized geometry in the sense of Muetterties [21].

Configuration index: The configuration index is a series of digits (the number of digits depends on the coordination polyhedron) identifying the relative positions of the ligating atoms on the vertices of the coordination polyhedron. Its meaning has to be specified for each type of polyhedron separately. The individual configuration index has the property that it distinguishes between diastereomers. The basis for the configuration index (and for some of the chirality descriptors, as will be discussed later) are *priority numbers* (called *seniority numbers* in Ref. 50), assigned to the ligating atoms of a mononuclear coordinating system according to the standard sequence rule developed for carbon compounds by Cahn, Ingold and Prelog [44, 45]. These rules are often referred to as the CIP rules and they are quoted below:

'The ligands associated with an element of chirality are ordered by comparing them at each step in bond-by-bond exploration of them, from the element, along the successive bonds of each ligand, and where the ligands branch, first along branch-pathways providing highest precedence to their respective ligands, the explorations being continued to total ordering by use of the standard Sub-Rules each to exhaustion in turn, namely:

(0) Nearer end of axis or side of plane precedes further.
(1) Higher atomic number precedes lower.
(2) Higher atomic mass number precedes lower.'

Four remarks have to be made in conjunction with the application of these CIP-rules to coordination compounds:

(i) The priority rules can be used in a generalized way, i.e. in principle for any coordination unit, not only if it represents an element of chirality. This is important for coordination compounds, where diastereomers are also described using ligand priorities.
(ii) Sub-Rule (0) is generally of no importance for mononuclear coordination entities.
(iii) Further existing CIP Sub-Rules are not needed for the present purposes.
(iv) In coordination entities (chiral or non-chiral), ligating atoms will have often assigned to them identical priority numbers. To distinguish these ligands, a higher priority is assigned to the ligating atom trans or opposite (on a structural axis) to the ligating atom of highest priority number.

Rule (1) is the basic instrument of the CIP system. It attributes priorities to ligands. We will examine a few practical cases for illustration:

(a) The priorities in the complex $[PtBrClFI]^{2-}$ are easily determined by the atomic mass of the ligands (4.14).

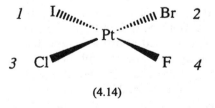

(4.14)

(b) In a TPY-3 molecule, the lone pair is considered to be a phantom ligand with mass zero. The priorities of the ligands in [PBrClF] are therefore as in (4.15), where italicized numbers indicate priorities of the ligands, including the phantom ligand.

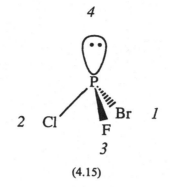

(4.15)

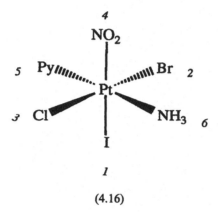

(4.16)

(c) The priorities of the ligands in the hexakis-heteroleptic complex of Pt^{IV} are given in (4.16).

In a case such as this, where identical ligand atoms are present (in this case three N atoms), the second coordination sphere is explored to determine the priorities of these ligands. Thus, NO_2 precedes pyridine, which in turn has a higher priority than NH_3. In more complicated cases one has to follow a tree graph (Figure 4.11) from the ligating atom 'outward,' where one explores each sphere completely. In this scheme the ligand spheres are depicted in such a way that the ligand of highest priority is on top of each branching. In the case of delocalized systems, phantom atoms (indicated as C000 and N00, respectively) have to be introduced such that a C000–N00 pair is added to a double bond. C000 becomes the new ligand of N and N00 of C. The priorities of the atom carrying the phantom atoms is assigned taking the average priority number from all possible valence structures. At each different branching the three different atoms are compared, one after the other beginning with the highest priority number. If no assignment can be made, the branch-tree is investigated further, beginning with the branch having the highest priority. If still no priority number can be assigned, the so called *trans*-maximum

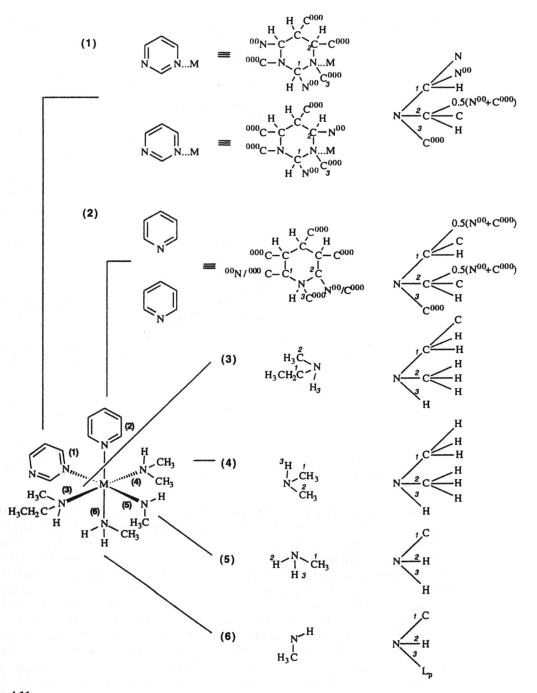

Figure 4.11
A 'tree graph' for the application of the CIP rules

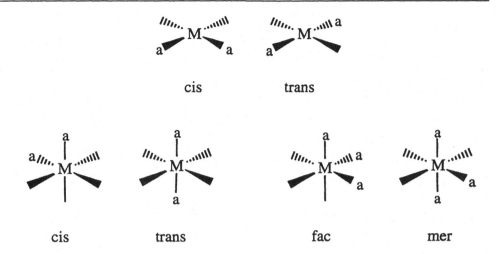

Figure 4.12
Configuration descriptors *cis/trans* and *fac/mer* for SP-4 and OC-6 complexes, respectively

principle applies, which states that the ligand *trans* to the highest priority number is considered prior to the ligand in the *cis* position.

With these rules, unique configuration indices of not more than three digits can be defined for CN ≤ 6.

Although alternative ways to distinguish diastereomers are no longer encouraged by the IUPAC Commission, they are still very popular among coordination chemists, and they will probably not disappear. These are *cis* and *trans* in SP-4 and OC-6 complexes, and *mer* (meridional) and *fac* (facial) in OC-6 complexes (Figure 4.12). Since these prefixes are very convenient, self-explanatory symbols, we will use them also when discussing appropriate structures.

Chirality symbols. The dichotomy of chiral objects requires, in principle, one dual reference system to which the enantiomeric structures can be matched by some unambiguous rules. From a practical point of view, however, it is useful to have reference systems that give some obvious relationship to the molecular structures under consideration. Three types of chirality symbols are used to designate chiral coordination entities: (i) symbols based on the priority rules, the so-called *steering-wheel* system, (ii) symbols based on the *skew-line* convention, and (iii) symbols based on the *oriented-lines* reference convention.

4.4.1 The Steering Wheel Reference System

The Cahn, Ingold, Prelog (CIP) rules were originally developed for carbon compounds containing stereogenic T-4 atoms [54]. The two symbols for the two enantiomers of a chiral pair are *R* and *S*, respectively, *R* being assigned to the enantiomer where the counting, starting with the ligand of highest priority, is clockwise, when the viewer is looking 'down' the vector from the carbon atom to the ligand of lowest priority. The converse applies for *S*. Figure 4.13 gives the convention for T-4, together with an example.

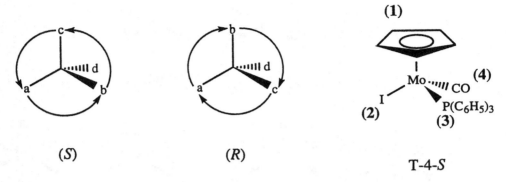

Figure 4.13
Steering-wheel for a T-4 complex. An example that has been resolved by Reisner *et al.* [55]

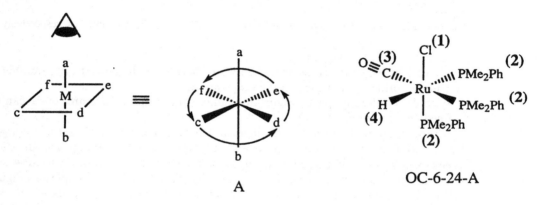

Figure 4.14
Steering-wheel for an OC-6 complex. The chirality symbol of the given example is *A*

Similar conventions can be formulated for other polyhedra. In order to avoid confusion, *R* and *S* are only applied for T-4, whereas in all other cases the symbols *C* (clockwise) and *A* (anti clockwise) are used. Figure 4.14 gives the convention together with an example.

It is important to note that *R/S* and *C/A* descriptors are purely formal designations, which bear no relation whatsoever to our every-day experience of left and right, and clockwise and anti-clockwise, respectively. The T-4 center (a) in Figure 4.15 has the *R* configuration, whereas (b) is *S* configured. No real inversion has occurred, however, when T was substituted by Br.

We call the transition that occurs from *R* to *S* in the case depicted in Figure 4.15 a *pseudo-inversion*. In a pseudo-inversion, the chirality descriptor changes, despite the fact that the transition is a simple substitution, where the incoming ligand occupies the same site as the substituted ligand. A pseudo-inversion is, in principle, always possible for substitutions at chiral T-4 centers.

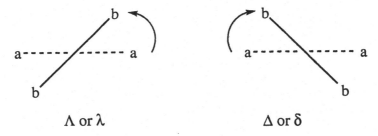

Figure 4.15
Pseudoinversion at a T-4 center. The descriptor changes from *R* to *S*, but the substitution has taken place with a simple exchange of one ligand only

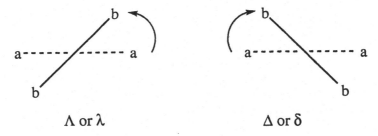

Λ or λ Δ or δ

Figure 4.16
A pair of *skew lines* and the definition of Δ/Λ and δ/λ

4.4.2 The Skew Line Reference System

A different convention can be formulated for any complex where two skew lines (Figure 4.16) define uniquely a helical system in a coordination entity. For this kind of chirality, descriptors are generally used and officially recommended by IUPAC that are not based on the CIP rule. Two (planar) chelates in the chiral configurations of Figure 4.16 can, e.g., be represented by a pair of skew lines.[†]

The two lines defining the chirality can be the projections of two (A═A) ligands, the $C \cdots C$ and the $N \cdots N$ connecting line in a (A≈A) ligand such as en. Such skew lines define the helix in OC-6 complexes (see Chapter 5, Section 5.3.1) with at least two bidentate ligands (leaving two ligand positions in a cis configuration occupied by other ligands), or in any five-membered, non-planar chelate ring. But also SP-4 complexes can be helically chiral if two planar bidentate ligands are not coplanar. Helical chirality can, of course, appear in many other CGs. The symbols for helical chirality of this type are Δ and Λ for chiralities involving the whole coordination unit and δ and λ for chiralities caused by conformers of chelate rings (see Chapter 5, Section 5.3.2). These chirality descriptors have a direct relation to well known macroscopic chiral objects (Δ and δ correspond to a right-handed screw), whereas the *R/S* and C/A descriptors are purely formal means of assigning a chirality to an object. Thus, there is no direct relation between the two nomenclatures, although the purely formal method can, in principle, always be applied.

[†]A pair of skew lines are two non-intersecting lines in space, which are neither parallel nor orthogonal to each other. Exactly one line, which intersects both lines, is orthogonal to both of them.

4.4.3 The Oriented Line Reference System

There are situations where a third reference system is more useful than either the steering wheel or the skew line conventions. It is called the oriented-(skew)-lines reference system [56]. In this system, two oriented lines define the chirality descriptors (Figure 4.17).

By analogy to the skew lines system, the symbols are chosen to be $\vec{\Delta}$ and $\vec{\Lambda}$, respectively. The chirality of the system is inverted if the orientation of *one* line is reversed, but unaltered if the orientation of both lines are inverted. Unlike the *skew-line* system, the *oriented-line* system is still chiral, when the two lines are mutually orthogonal. Upon rotation of one line around the mutually orthogonal axis connecting the two lines, the chirality changes sign with a periodicity of 2π for the oriented line system and a periodicity of π in the skew-line system. The *oriented-line* system has always one C_2 symmetry axis (symmetry group C_2), except for the anti-parallel orientation, where in addition a center of inversion and a mirror plane occurs (symmetry group C_{2h}).

We will use the oriented line system in cases where the skew line system is not applicable, either because the two lines are orthogonal (but the molecule is still chiral), or because it gives an ambiguous result.

Other stereochemical concepts: In organic chemistry several new concepts in stereochemistry concerning *prostereoisomerism* have evolved during the past 30 years. They have proven to be of great value in many respects. One of the driving forces of this evolution was undoubtedly the enormous progress in NMR spectroscopy and the wide applicability of this technique. Coordination chemistry overlaps in a broad range with organic chemistry, and these concepts are therefore relevant to the development of the field, as far as stereochemical characterization of organic ligands is concerned. In the field of stereochemistry of elements with CN > 4, which is of central interest for coordination chemistry but which is outside the scope of traditional organic chemistry, the application of analogous concepts is just beginning to emerge [48]. A comprehensive, up-to-date presentation of the historic development and of the state of this field in *organic chemistry* is given by Eliel and Wilen [15, Chapter 8]. The interested reader is referred to the two publications cited. Here, we will discuss only the most basic concepts in view of their application to coordination compounds.

The basis for the prostereoisomerism was already given in Section 4.3.2. There are several concepts that need to be introduced in this connection. Two ligands that are identical, if detached from the coordination center are called *homomorphic*.

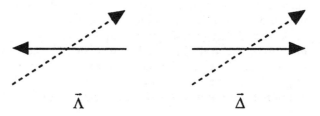

$\bar{\Lambda}$ $\bar{\Delta}$

Figure 4.17
A pair of *oriented lines* and the definition of $\vec{\Lambda}$ and $\vec{\Delta}$

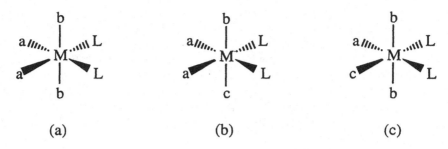

Figure 4.18
Three different types of a pair of *homomorphic* ligands (L)$_2$: they are *homotopic* (related by a proper rotation, C_2 in this case) in (a) and *heterotopic* in (b) and (c); in (b) they are *enantiotopic* (related by an improper rotation, C_s in this case) and in (c) *diastereotopic* (not related by any symmetry element)

Homomorphic ligands may be *homotopic or hetereotopic*, and heterotopic ligands may be *enantiotopic or diastereotopic*. Homotopic/heterotopic and enantiotopic/ diastereotopic can be applied not only to ligands, but also to faces of planar molecules. This is equivalent to looking at faces of a planar molecule as being occupied by two ligands of mass zero, which are often called *phantom* ligands. Phantom ligands are obviously homomorphic. We can introduce here these notions in a purely topographical way, using OC-6 complexes as bases for the definitions (Figure 4.18).

We are looking at the complexes OC-6 [Ma$_2$b$_2$L$_2$] (a), [Ma$_2$bcL$_2$] (b), and [Mab$_2$cL$_2$] (c) in (Figure 4.18). We focus our attention to the two ligands L. They are obviously homomorphic. In (a) and (b), respectively, they are equivalent in the sense that they are interchanged by symmetry operations. In (a), which belongs to the symmetry group C_{2v}, the two ligands L are interchanged by a rotation about the C_2 axis, bisecting the L–M–L angle *and* by the reflection at the mirror plane, which is perpendicular to the Ma$_2$L$_2$ plane. In (b), which belongs to C_s, the L ligands are interchanged by the only mirror plane of the molecule, whereas in (c), the two L ligands lie in the plane of the C_s-symmetric molecule, but they are not transformed into each other by any symmetry operation of the molecule. Substitution of one or the other of the two L ligands by a new ligand L' leads to *two identical* molecules in the case of (a), to a *pair of enantiomers* in the case of (b), and to *two diastereomers* in the case of (c). The two ligands are called *homotopic* in the case of (a) and *heterotopic* in the cases of (b) and (c). They are *enantiotopic* in (b) and *diastereotopic* in (c). The symmetry criteria given above can be generalized:

- Ligands are *homotopic* if they interchange positions through operation of a proper rotation symmetry axis.
- Ligands are *enantiotopic* if they exchange positions through operation of an improper rotation axis. They must not exchange positions if a rotation is carried out around a proper rotation axis, which may otherwise be present in the molecule.
- Ligands that are *diastereotopic* must not be related either by a proper or by an improper rotation symmetry element.

Since in the case of complex (b), a chiral complex occurs through substitution of one of two homomorphic, enantiotopic ligands in a achiral complex, the former is called prochiral. Prochirality of a molecule requires enantiotopic ligands (including phantom ligands), which in turn require a symmetry element of improper rotation. Two consequences result from these considerations: (a) a molecule can be either prochiral or chiral, but it cannot have both properties simultaneously, and (b) heterotopic ligands in a chiral molecule are necessarily diastereotopic. We will discuss some prochiral complexes later, when the corresponding stereoisomers are considered.

4.5   Stereochemical Language

There is a distinct difference between nomenclature and language in a given field of science. The former represents a set of rules elaborated by some international committee (in chemistry the generally accepted body is IUPAC) for naming subjects of that field of science. In the case of coordination chemistry, which penetrates many fields of chemistry, and which is especially closely related to organic chemistry, the problem of setting up these rules is a formidable one. A well defined nomenclature is especially important in today's chemistry, since it is a way for data retrieval by computer from the enormous existing databases. Some journals insist more, others less, strictly on the use of official nomenclature in their publications. It is, however, a fact of life that it is impossible to use official nomenclature always, and especially not in everyday practical language. Names for compounds would become so extremely complicated that *ad hoc* abbreviations are unavoidable and in fact necessary.

Chemistry in general has developed to a state where nomenclature in the traditional sense is often no longer useful, since the names become much too complicated in many practical cases. It would be highly desirable if a radical new approach, i.e. automatic computer coding from structural formulae, could be established. There is little doubt that this method will replace traditional chemical nomenclature sometimes in the 21st century.

An alternative way for identification of a chemical compound or of a type of molecule is to number it instead of naming it, a system pursued by Chemical Abstracts Service, with the CAS number. This might be the most convenient way of identification, since with a simple number all information can be obtained from data bases using the numbering system. The problem of identifying the number for a given molecule remains, however, and for this nomenclature is unavoidable.

Another problem is the usage of terms. Since science evolves, the terms used to describe facts have to evolve too. Some, which seem to have been firmly established for a long duration, suddenly lose their solid foundation and have consequently to be replaced by new, better defined terms. A case in point is asymmetric or chiral atom. Since Pasteur, asymmetric was used mainly in organic chemistry for the case of a carbon atom with four different ligands. But asymmetric was also used by Werner for OC-6 Co-complexes showing chirality (Spiegelbildisomerie). Currently *stereogenic* center is preferred over *asymmetric carbon atom* for various, justified reasons [41]. As mentioned earlier, other concepts such as 'geometrical' or 'dissymmetric' may outgrow their usage. Others, such as 'catoptromers' or 'allogons', even though logical and in principle useful, have never found the acceptance of a large part of the scientific community which is involved.

In order to cope with these problems, Appendix III gives common abbreviations of ligands together with their structural formulas, and Appendix II gives a glossary of terms in stereochemistry, as they are used in this book.

4.6 References

1. Cotton, F. A., *Chemical Applications of Group Theory*, 2nd edn, Wiley, Chichester, 1971.
2. Bethe, H. (1929), *Ann. Phys.*, **3**, 133–208.
3. Atkins, P. W., Child, M. S. and Phillips, C. S. G., *Tables for Group Theory*, Oxford University Press, New York, 1990.
4. Anet, F. A. L., Miura, S. S., Siegel, J. and Mislow, K. (1983), *J. Am. Chem. Soc.*, **105**, 1419–1426.
5. Flurry, R. L., Jr (1981), *J. Am. Chem. Soc.*, **103**, 2901–2902.
6. Jørgensen, C. K., *Oxidation Numbers and Oxidation States*, Springer, Berlin, 1969.
7. Schauer, C. K. and Anderson, O. P. (1988), *Inorg. Chem.*, **27**, 3118–3130.
8. Harrowfield, J. M. and Wild, S. B., in *Comprehensive Coordination Chemistry*, G. W. Wilkinson (Ed.), Vol.1, Pergamon Press, Oxford, 1987, pp. 179–212.
9. Buckingham, D. A., *Structure and Stereochemistry of Coordination Compounds*, in *Inorganic Biochemistry*, Vol. 1, Elsevier, Amsterdam, 1973, pp. 3–62.
10. Cornioley-Deuschel, C. and Von Zelewsky, A. (1987), *Inorg. Chem.*, **26**, 3354–3358.
11. Burmeister, J. L. (1968), *Coord. Chem. Rev.*, **3**, 225–245.
12. Burmeister, J. L., in *The Chemistry and Biochemistry of Thiocyanate Acid and Its Derivatives*, A. A. Newman (Ed.), Academic Press, London, 1975.
13. Prelog, V., in *ACS Symposium Series*, B. Ramsey (Ed.), Vol. 12, American Chemical Society, Washington, DC, 1975, pp. 179–188.
14. Buckingham, D. A., Maxwell, I. E. and Sargeson, A. M. (1969), *J. Chem. Soc., Chem. Commun.*, 581–583.
15. Eliel, E. L. and Wilen, S. H., *Stereochemistry of Organic Compounds*, Wiley–Interscience, New York, 1994.
16. Barron, L. D. (1986), *Chem. Phys. Lett.*, **123**, 423–427.
17. Pasteur, L. (1848), *Ann. Chim.*, **24**, 459–460.
18. Saito, Y. (1978), *Top. Stereochem.*, **10**, 95–174.
19. Saito, Y., *Inorganic Molecular Dissymmetry, Inorganic Chemistry Concepts*, Vol. 4, Springer, Berlin, 1979.
20. Lord Kelvin, in *Baltimore Lectures*, Cambridge University Press, Cambridge, 1904.
21. Muetterties, E. L. (1970), *Acc. Chem. Res.*, **3**, 266–273.
22. Kilbourn, B. T., Powell, H. M. and Darbyshire, J. A. C. (1963), *Proc. Chem. Soc.*, 207–208.
23. Raymond, K. N., Corfield, P. W. R. and Ibers, J. A. (1968), *Inorg. Chem.*, **7**, 1362–1372.
24. De Camp, W. H. (1989), *Chirality*, **1**, 2.
25. Halevi, E. A. (1992), *Chem. Eng. News*, **70**, 2.
26. Stinson, S. C. (1992), *Chem. Eng. News*, **70**, 46–76.
27. Rotheim, P. (1992), *Chem. Eng. News*, **70**, 3.
28. Collet, J.-J. A. and Wilen, S. H., *Enantiomers, Racemates and Resolutions*, Wiley, New York, 1981.
29. Bernal, I. (1992), *J. Chem. Educ.*, **69**, 468–469.
30. Bernal, I., Cai, J. and Myrczek, J. (1993), *Polyhedron*, **12**, 1157–1162.
31. Bernal, I., Myrczek, J. and Cai, J. (1993), *Polyhedron*, **12**, 1149–1155.
32. Bernal, I. and Kauffman, G. B. (1987), *J. Chem. Educ.*, **64**, 604–610.
33. Bernal, I. and Kauffman, G. B. (1993), *Struct. Chem.*, **4**, 131–138.
34. Bernal, I., Cetrullo, J., Myrczek, J., Cai, J. and Jordan, W. T. (1993), *J. Chem. Soc., Dalton Trans.*, 1771–1776.

35. Haupt, H. J. and Huber, F. (1978), *Z. Anorg. Allg. Chem.*, **442**, 31–40.
36. Albano, V. G., Bellon, P. and Sansoni, M. (1971), *J. Chem. Soc. A*, 2420–2425.
37. Albano, V. G., Bellon, P. L. and Sansoni, M. (1969), *J. Chem. Soc., Chem. Commun.*, 899–901.
38. Albano, V. G., Ricci, G. M. B. and Bellon, P. L. (1969), *Inorg. Chem.*, **8**, 2109–2115.
39. Barron, L. D. (1986), *J. Am. Chem. Soc.*, **108**, 5539–5542.
40. Werner, A. and Vilmos, A. (1899), *Z. Anorg. Allg. Chem.*, **21**, 145–164.
41. Mislow, K. and Siegel, J. (1984), *J. Am. Chem. Soc.*, **106**, 3319–3328.
42. Macomber, R. S. (1994), *Chem. Eng. News*, **72**, 2.
43. McCasland, G. E., Horvat, R. and Roth, M. R. (1959), *J. Am. Chem. Soc.*, **81**, 2399–2402.
44. Cahn, R. S., Ingold, C. and Prelog, V. (1966), *Angew. Chem., Int. Ed. Engl.*, **5**, 385–415.
45. Prelog, V. and Helmchen, G. (1982), *Angew. Chem., Int. Ed. Engl.*, **21**, 567–583.
46. Mills, W. H. and Quibell, T. H. H. (1935), *J. Chem Soc.*, 839–846.
47. Hanson, K. R. (1966), *J. Am. Chem. Soc.*, **88**, 2731–2742.
48. Mestroni, G., Alessio, E., Zassinovich, G. and Marzilli, L. G. (1991), *Comments Inorg. Chem.*, **12**, 67–91.
49. Leigh, G. J., *Nomenclature of Inorganic Chemistry*, Blackwell, Oxford, 1990.
50. Block, B. P., Powell, W. H. and Fernelius, W. C., *Nomenclature of Inorganic Chemistry–Recommendations*, ACS Professionnal Reference Book, American Chemical Society, Washington, DC, 1990.
51. Brown, M. F., Cook, B. R. and Sloan, T. E. (1975), *Inorg. Chem.*, **14**, 1273–1278.
52. Brown, M. F., Cook, B. R. and Sloan, T. E. (1978), *Inorg. Chem.*, **17**, 1563–1568.
53. Sloan, T. E., in *Topics in Inorganic and Organometallic Stereochemistry*, G. L. Geoffrey (Ed.), Vol. 12, Wiley, New York, 1981, pp. 1–36.
54. Cahn, R. S., Ingold, C. and Prelog, V. (1956), *Experientia*, **12**, 81–94.
55. Reisner, G. M., Bernal, I., Brunner, H. and Muschiol, M. (1978), *Inorg. Chem.*, **17**, 783–789.
56. Damhus, T. and Schäffer, C. E. (1983), *Inorg. Chem.*, **22**, 2406–2412.

5 Topographical Stereochemistry of Mononuclear Coordination Units

5.1 Coordination Units with Achiral, Monodentate Ligands Only. Coordination Polyhedra and Isomers

Coordination Numbers 2 and 3

The coordination sites with polyhedral symbols L-2, A-2, and TP-3 do not give rise to any remarkable topographical stereochemistry and they will therefore not be discussed any further.

TPY-3. This case is actually rare for metallic coordination centers. It is, however, very important for the non-metallic elements of Group 15. From a topographic stereochemical point of view, it is closely related, in fact isomorphous, to T-4 sites. The lone pair that usually is responsible for a molecule rendering CN = 3 to be non-planar can be regarded as a special kind of a ligand, often called a 'phantom' ligand. TPY-3 can therefore be treated as T-4 with the *lp* being the 'ligand' with the lowest priority number.

Coordination Number 4

T-4. As is well known, from the three elements which are in general necessary to characterize a coordination unit from the topographical point of view, only two are needed for T-4: the polyhedral symbol itself and a chirality descriptor, in the case that all four ligands are different. The configuration index is not needed, since a mononuclear T-4 center cannot give rise to diastereomers. The chirality descriptors are *R* and *S*, respectively, where the assignment is given by the CIP convention. The view is from M along the pseudo-threefold axis toward the bonding atom with the lowest priority (d). If the three ligands a, b, c (ordered according to their priorities) are arranged in a clockwise way, the configuration is *R*, otherwise *S*.

As mentioned already, metal T-4 centers are in almost all cases (with the exception of the pseudo-tetrahedral cp complexes discussed below) substitutionally labile, whereas carbon T-4 centers are almost always substitutionally inert. The rich stereochemistry of organic compounds is therefore not known for metal centers of this geometry. The lability of metal complex T-4 centers, on the other hand, gives rise to polynuclear assemblies with interesting stereochemical properties. Such complexes will be discussed later.

No 'asymmetric' coordination unit T-4 [Mabcd] was known until 1969, when Brunner prepared the first such complex (Figure 5.1), in which the cyclopentadienyl ligand is considered to be monodentate [1–4].

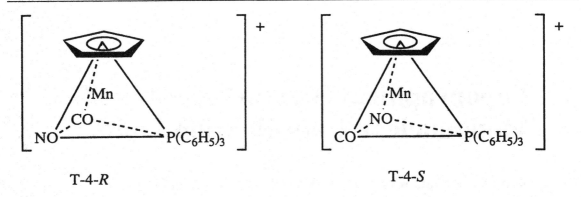

Figure 5.1
The enantiomeric pair of T4 $[Mn(cp)(P(C_6H_5)_3)(CO)(NO)]^+$

Such complexes do not racemize under mild conditions, and they can undergo many reactions, some of them will be discussed later. A different way of looking at these complexes is to consider them as octahedral, where the cp ligand occupies one triangular face of the octahedron, and the other three ligands the opposite face [5,6]. We do not consider this question further here, because organometallic compounds are in general outside the scope of this book.

SP-4. The other CG with CN = 4 is SP-4, unknown in carbon chemistry and very rare in coordination units of Main Group elements, but not uncommon among transition elements. It is the predominant CG for the d^8 metals Ni^{II} (only with strong field ligands), Pd^{II}, Pt^{II}, Co^I, Rh^I, Ir^I, Cu^{III}, Ag^{III}, and Au^{III}, and it appears also for l.s. d^7 in Co^{II} and for d^9 in Cu^{II} and Ag^{II}. A large part of the interest in the chemistry of Pt^{II} complexes is focused on stereochemical properties, which are closely connected with the *trans*-influence series.

Table 5.1 gives all possible SP-4 complexes with monodentate ligands.

As opposed to T-4, in a general case SP-4 requires for a complete stereochemical

Table 5.1. Number of isomers and symmetries of SP-4 complexes with monodentate ligands

Complex type	Total number of isomers	Symmetries
Homoleptic [Ma_4]	1	D_{4h}
Bis-heteroleptic [Ma_3b] [Ma_2b_2]	1 2	C_{2v} D_{2d}, C_{2v}
Tris-heteroleptic [Ma_2bc]	2	C_{2v}, C_s^a
Tetrap-heteroleptic [Mabcd]	3	C_s^a, C_s^a, C_s^a

[a]Prochiral complexes (see text).

characterization no chirality descriptor (at least not in coordination units with monodentate ligands), but a configuration index (or a configuration descriptor, *cis* or *trans*) is needed. The configuration index for SP-4 is a one-digit number, given by the priority ranking number of the ligand *trans* to the ligand with the highest priority. With two different ligands, two diastereomers (often called geometrical isomers in this case) are possible for [Ma₂b₂]. In this case, the configuration descriptors *cis* and *trans* are unambiguous. As an example, the complexes already mentioned by Werner are given in (5.1).

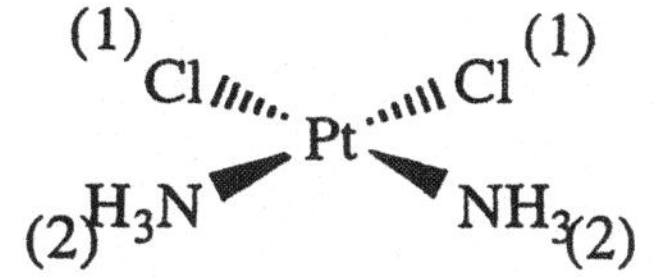

cis-[PtCl₂(NH₃)₂]
SP-4-2

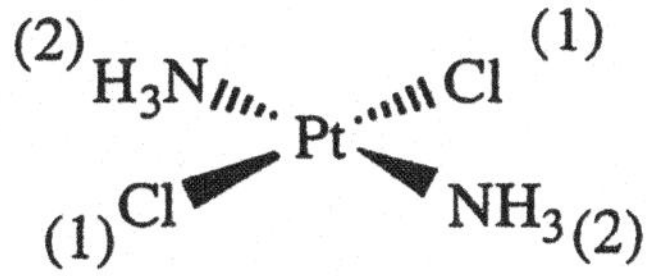

trans-[PtCl₂(NH₃)₂]
SP-4-1

(5.1)

With three different ligands, again two diastereomers are possible (5.2), and they can be designated as *cis* and *trans* (with respect to the two homomorphic ligands), or by the configuration index.

With four different ligands (5.3), three diastereomers are possible and the designation with *cis* and *trans* evidently does not suffice to designate them. For the classical case of [Pt(NH₃)(NH₂OH)(NO₂)(py)], for which all possible diastereomers were synthesized [7,8], the stereochemical descriptors are given in (5.3).

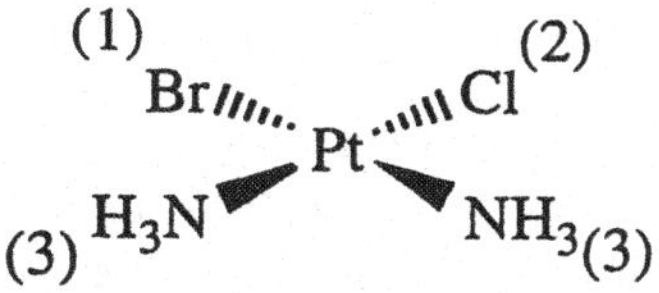

cis-[PtClBr(NH₃)₂]
SP-4-3

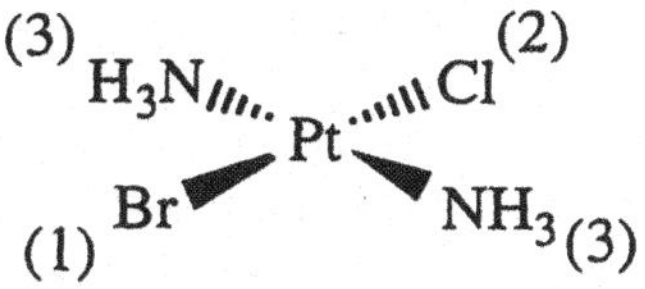

trans-[PtClBr(NH₃)₂]
SP-4-2

(5.2)

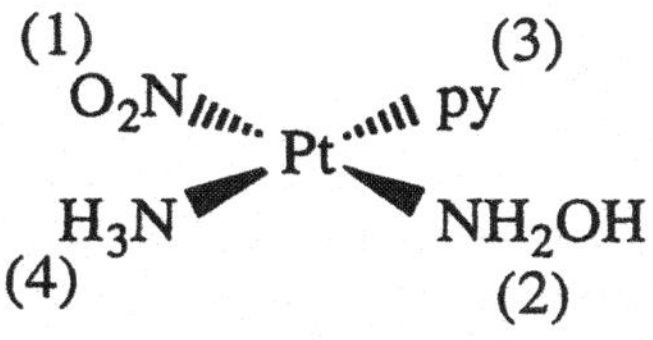

SP-4-2

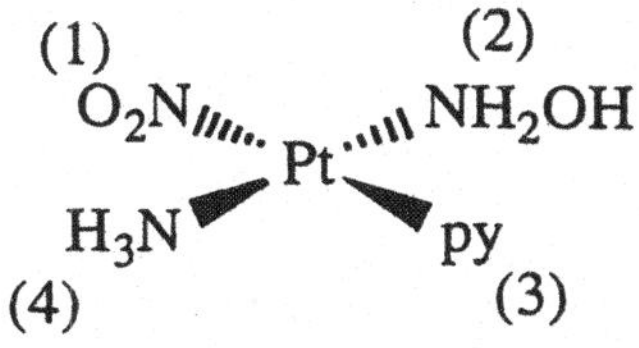

SP-4-3

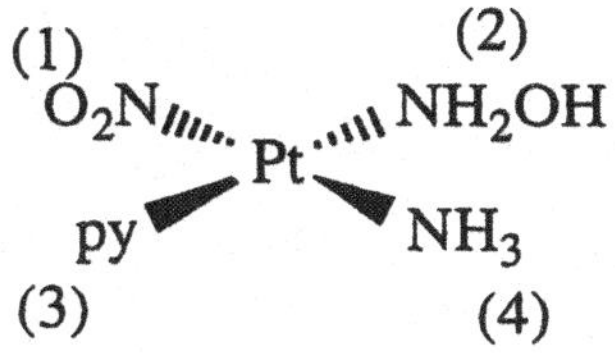

SP-4-4

(5.3)

An important family of SP-4 complexes are the so-called Vaska's compounds [9] [Ir(CO)(Cl)(PR$_3$)$_2$] SP-4-3, i.e. *trans* (5.4). They will be discussed in Chapter 7 in more detail as prototypes of complexes undergoing oxidative addition reactions.

$$
\begin{array}{cc}
(2) & (1) \\
R_3P\cdots & \cdots Cl \\
& Ir \\
OC & PR_3 \\
(3) & (2)
\end{array}
$$

SP-4-3

(5.4)

As indicated in Table 5.1, complexes of the type SP-4 *cis*-[Ma$_2$bc] and SP-4 [Mabcd] are prochiral with respect to *trans* addition of two ligands forming an OC-6 complex. They have enantiotopic faces. Note that the Vaska complex is not *prochiral*.

Relationships between T-4 and SP-4. Although both coordination geometries for coordination number four appear frequently in coordination units of transition metals, *polytopal isomerism* occurs rarely, but several examples are well documented. Ni^{II} complexes, where the average ligand field strength is not very large and where steric interactions between ligands are important, seem to provide the best basis for the observation of T-4/SP-4 polytopism. Besides examples with chelate ligands, which will be discussed later, complexes with monodentate ligands are known that exhibit polytopal isomerism. [NiX$_2$(PR$_3$)$_2$] compounds exist as T-4 centers with X = Br$^-$ or I$^-$, and R = aryl, but SP-4 is preferred with X = Cl or NCS$^-$, and R = alkyl [10]. In certain cases, as in [NiBr$_2$(PBzPh$_2$)$_2$], both geometries can co-exist in the same solid [11]. The latter compound contains three molecules per unit cell, one having *trans*-SP-4 configuration, and two T-4 coordination geometry. Except for these rare cases, the conditions for the occurrence of the two geometries are mutually exclusive. SP-4 can occur only if electronic factors (MOSE in the case of transition metals and two lone pairs in a six valence state electron pair molecule) are present, which stabilize the SP arrangement strongly. For this reason, SP-4 complexes are generally diastereomerically stable; even so, isomerization could occur via a T-4 transition state, as depicted in Figure 5.2.

Coordination Number 5

In contrast to CN 4, where for both coordination geometries configurationally rigid and substitutionally inert coordination units exist (although the latter condition is for T-4 only fulfilled by the carbon atom), stereoisomerism generally does not occur with CN5 in complexes with monodentate ligands. Heteroleptic coordination centers with monodentate ligands are rare. Isolable stereoisomers have been reported, however. Two cases where diastereomers could be separated are SPY-5

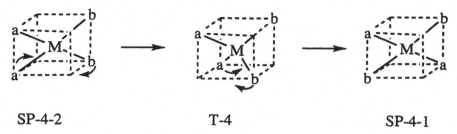

Figure 5.2
Cis/trans isomerization of SP-4 through T-4

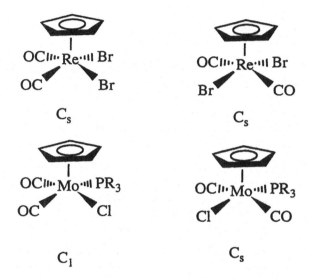

Figure 5.3
Diastereoisomers of Re(III) and Mo(II) complexes with SPY-5 which have been isolated

complexes of Re^{III} [12], and Mo^{II} [13–15], respectively, where cp$^-$ occupies the apical position (Figure 5.3). Here again, cp$^-$ is considered to be a monodentate ligand. This is practical for stereochemical purposes, whereas for discussions of bonding properties, the 6e-pentahapto coordination of this ligand has to be taken explicitly into consideration.

In the nomenclature scheme of IUPAC 1990, conventions are defined for stereochemical descriptors in TB-5 (= TBPY-5) (Figure 5.4). These conventions are the following:

- *Orientation*: View from axial ligand with higher priority toward M.
- *Configuration index*: Two-digit number, which gives the ranking of the two axial ligands in numerical order.
- *Chirality symbols*: *C* or *A* according to the ranking order of equatorial ligand priorities.

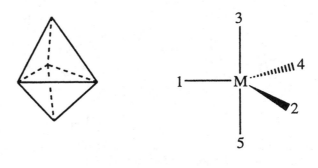

TB-5-35

Figure 5.4
Configuration index for TB-5

Note that at least four ligands have to be different in order that a TB-5 complex can be chiral; however, there are complexes with four different ligands that are not chiral, namely those where the chirality index is a number with two equal digits. Prochirality of some five coordinate complexes was discussed by Mestroni *et al.* [16].

In SPY-5 coordination, similar conventions apply (Figure 5.5). The convention is defined in the following way:

- *Orientation*: View from the apex of the pyramid to M.
- *Configuration index*: Priority ranking number of the apex ligand, followed by the priority ranking number of the ligand *trans* to the highest priority ligand in the base of the square pyramid.
- *Chirality symbols*: *C* or *A*, for at least four different ligands.

Note that SPY-5 can be stereochemically treated as OC-6, with one phantom ligand (the lacking ligand from OC-6) always having the lowest priority.

The complexes depicted in Figure 5.3 are SPY-5 only under the assumption that cp⁻ occupies only one coordination site, i.e. if it is considered to be a monodentate ligand. From a symmetry point of view, this is justified since it is known that the energy barrier for the rotation of the cp ring is very low, and its rotation

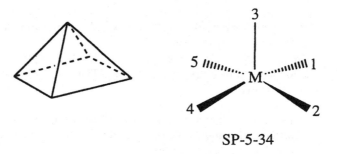

SP-5-34

Figure 5.5
Configuration index for SPY5

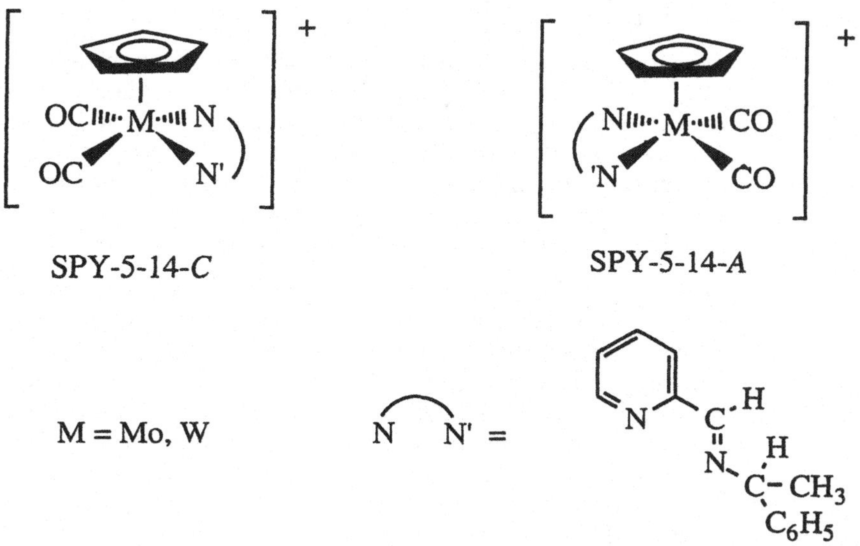

Figure 5.6
Chiral SPY-5 complexes, where the enantiomers have been separated

consequently rapid. Taking this into account, the symmetries are those given in Figure 5.3. The asymmetric *cis*-molybdenum complex is the only chiral one in this series. It was separated into the enantiomers [17] by using an enantiomerically pure chiral *S*-aminophosphine. Therewith the two 'enantiomers' *C* and *A*, respectively, became diastereomers *CS* and *AS*.

Other five-coordinated species for which the enantiomers were separated are complexes with chelate ligands. Examples are given in Figure 5.6.

Relationships between TB-5 and SPY-5. Contrary to the case of CN 4, the two geometries of CN 5 are in general energetically close and consequently interconvertible. This gives rise to the well known stereochemical flexibility in five-coordination. Consequently, polytopal isomers are known. The best documented case is $[Cr(en)_3][Ni(CN)_5]1 \cdot 5H_2O$, which has been shown to have both coordination geometries present in the crystal [18]. Whether this polytopal isomerism exists in solution is not known. The complex $[Ni(CN)_5]^{3-}$ is thermodynamically highly unstable in solution (the equilibrium constant for the binding of the fifth CN^- ligand is small), and even with a high concentration of cyanide ion in solution, the relative concentration of the pentacoordinated complex is low. The presence of the pentacoordinated complex in the crystal lattice is possible because of the large lattice energy in an ionic solid with a cation of 3+ charge.

In five coordinated d^0-complexes, a TB-5 geometry would be expected to be more stable than the SPY-5 arrangement. As will be discussed for coordination number six below, the VSEPR model seems to break down under special circumstances. An experimentally well documented case is $[Ta(CH_3)_5]$, which has been found to be SPY-5 in the gas phase [19].

Coordination Number 6

What coordination number four signifies for organic chemistry, so does coordination number six for coordination chemistry. With six ligands in the rigid, and often substitutionally inert octahedral arrangement, an enormously rich stereochemistry becomes possible. Even 100 years after its introduction into chemistry, the octahedral coordination geometry is far from being completely explored from an experimental point of view. Two coordination polyhedra have to be considered, the octahedron OC-6 and the trigonal prism TP-6. The former is by far the more frequent.

OC-6. The complete stereochemical descriptor needs a two-digit configuration index, and a chirality symbol for chiral complexes. The view is defined from the ligand with highest priority to the metal and to the ligand with lower priority ranking, if there is a choice (Figure 5.7). The configuration index is the priority ranking number of the ligand *trans* to the ligand having highest priority, followed by the priority ranking number of the ligand *trans* to the ligand with highest priority[†] in the square perpendicular to the viewing axis.

For general discussions, locant descriptors *cis* and *trans* and *mer* and *fac* can be used if the designation is unambiguous, but they are not recommended for nomenclature purposes by IUPAC. Examples are given in (5.5) and in (5.6) (the IUPAC stereochemical descriptors are also given for priority ranking numbers according to alphabetical ordering of ligands).

Chirality symbols. *C* and *A* are defined with respect to the priority ranking sequence in the plane perpendicular to the viewing direction. *C* and *A* are generally used as chirality symbols in coordination units with monodentate ligands. In coordination units with chelate ligands, an alternative pair of chirality descriptors (Δ and Λ) will be introduced.

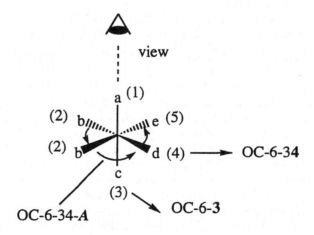

Figure 5.7
Stereochemical descriptors for OC-6

[†]One has to be careful not to confuse 'priority' and 'priority number.' The highest priority corresponds to the lowest priority number and vice versa.

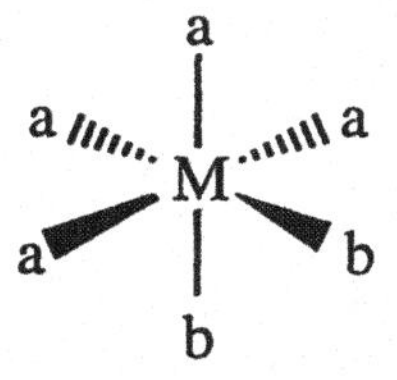

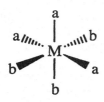

cis-[Ma$_4$b$_2$]
OC-6-22-[Ma$_4$b$_2$]
Highest possible symmetry: C$_{2v}$

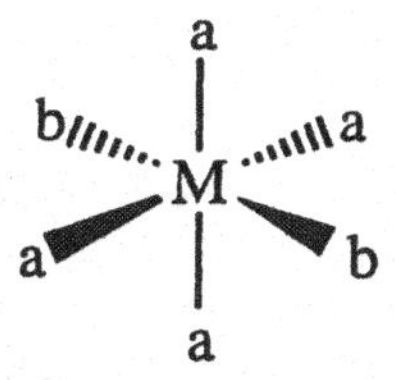

trans-[Ma$_4$b$_2$]
OC-6-11-[Ma$_4$b$_2$]
Highest possible symmetry: D$_{4h}$

(5.5)

mer-[Ma$_3$b$_3$]
OC-6-21-[Ma$_3$b$_3$]
Highest possible symmetry: C$_{2v}$

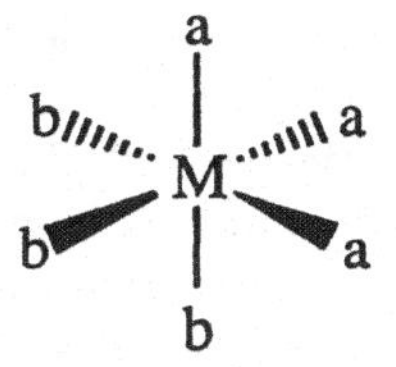

fac-[Ma$_3$b$_3$]
OC-6-22-[Ma$_3$b$_3$]
Highest possible symmetry: C$_{3v}$

(5.6)

Enumeration of the diastereomers of OC-6 coordination units with monodentate ligands.
There is a series of eleven possible compositions of complexes (for the use of
multiplicative prefixes see Ref. 20, p. 63):

- homoleptic   [Ma$_6$]
- bis-heteroleptic   [Ma$_5$b]; [Ma$_4$b$_2$]; [Ma$_3$b$_3$]
- tris-heteroleptic   [Ma$_4$bc]; [Ma$_3$b$_2$c]; [Ma$_2$b$_2$c$_2$]
- tetrakis-heteroleptic   [Ma$_3$bcd]; [Ma$_2$b$_2$cd]
- pentakis-heteroleptic   [Ma$_2$bcde]
- hexakis-heteroleptic   [Mabcdef]

The series cotains a total of 75 stereoisomers. Although there seems to be only one
documented report of the hexakis-heteroleptic case [21,22], the consideration of the
diastereomers of this most general case yields a useful basis for all OC-6
coordination units.

A hexakis-heteroleptic[†] OC-6 atom gives 15 pairs of enantiomers, or a total of 30
stereoisomers. Following a proposition by Bailar [23], the 15 diastereomers can be

---

[†]This is to be preferred over 'asymmetric,' which is analogous to the 'asymmetric' carbon atom, but
otherwise not very informative. All configurations of the hexa-heteroleptic complex are of symmetry C$_1$.
This symmetry occurs already for a tris-heteroleptic complex [Ma$_2$b$_2$c$_2$] OC-6-41, which can accordingly
be C or A.

represented in a table, the so-called Bailar Tableau, which can be constructed as follows:

- The position of ligand 'a' is invariant.
- A pair of letters signifies two ligands in *trans* position.
- In the first row of the table, the *trans* pair 'ab' is invariant.
- The three possibilities for the two other pairs are obtained by exchange of 'd' with 'e' and of 'd' with 'f,' respectively. In this way the three possibilities 1L, 1M, and 1N are obtained.

Then, 'c' is taken *trans* to 'a,' which yields row 2, etc. A total of $3 \times 5 = 15$ isomers are obtained in this way.

The Bailar Tableau (Table 5.2) gives the 15 diastereomeric hexa-heteroleptic complexes [Mabcdef]. The two-digit configuration indices are given below the representations in the stereo-centers (Figure 5.8). All configurations are chiral due to their C_1 symmetry, i.e. there is always a non-congruent mirror image molecule with opposite chirality, bringing the total number of stereoisomers of complexes of this type to 30. The chirality descriptors of all complexes given above is *A*. As pointed out by Mayper [24], chiral structures yield a non-identical symbol, namely that of the enantiomer with opposite configuration, after any odd number of interchange operations, and an identical symbol after any even number of interchange operations (an interchange is either the reversal of one *trans* pair, or the exchange of two *trans* pairs). The 15 configurations with *C*-chirality can therefore be generated very easily by inverting all the pairs with the a ligands, leaving everything else unchanged.

The five diastereomers so far synthesized of one hexakis-heteroleptic complex, [PtBrClINH$_3$NO$_2$py], by Essen and co-workers [21,22] are given in Figure 5.9.

Table 5.2. Bailar Tableau giving all 15 diastereomers for an OC-6 hexakis-heteroleptic complex [Mabcdef]

	L	M	N
1	ab	ab	ab
	cd	ce	cf
	ef	df	de
2	ac	ac	ac
	bd	be	bf
	ef	df	de
3	ad	ad	ad
	bc	be	bf
	ef	cf	ce
4	ae	ae	ae
	bc	bd	bf
	df	cf	cd
5	af	af	af
	bc	bd	be
	de	ce	cd

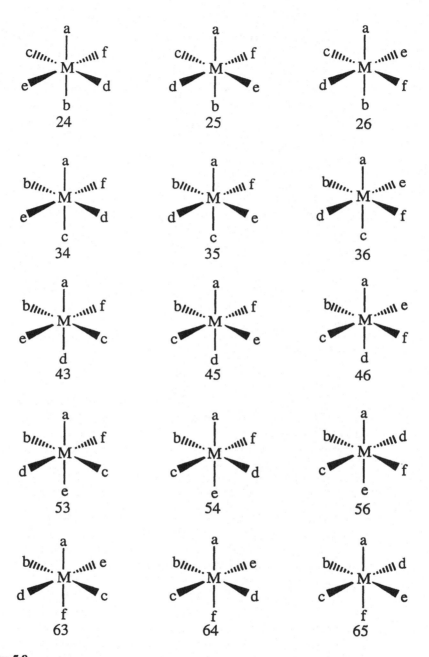

Figure 5.8
Representation of the 15 diastereomers of an [Mabcdef] complex, with the corresponding configuration indices. The absolute configuration for all representations is *A*

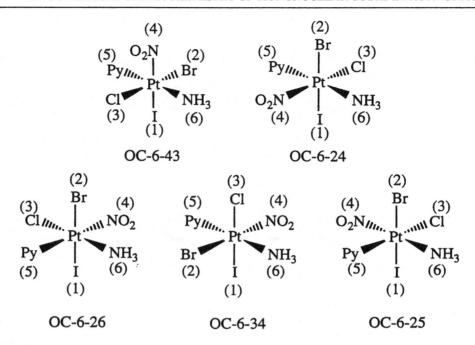

Figure 5.9
Diastereomers of the hexakisheteroleptic complex [PtBrClINH$_3$NO$_2$Py] prepared by Essen and co-workers [21,22]

All more simple cases with monodentate ligands can be obtained from the basic Bailar Tableau. As an example, let us examine [M(a)$_2$(b)$_2$(c)$_2$]. For the changes in the Bailar Tableau, a word processing computer program is very useful for constructing a new table. One can proceed in the following way:

- <Change f to a; change e to b; change d to c>.
- Then look for multiple entries, and for enantiomers.

Five entries in the Bailar Tableau for [Ma$_2$b$_2$c$_2$] (Table 5.3) are unique. Only 1L is transformed into a non-identical symbol by an odd number of interchanges, and it is consequently the only chiral isomer of this composition. A total of six stereoisomers, five diastereomers, and one pair of enantiomers is thus obtained for this type of complex (Figure 5.10).

An early realized example of a complex of this composition is [Pt(NH$_3$)$_2$(py)$_2$Cl$_2$]$^{2+}$ [8]. In 1979 two purely inorganic complexes of the *cis,cis,cis*-[Ma$_2$b$_2$c$_2$] type were resolved into the enantiomeric forms, namely *cis,cis,cis*-[Co(NH$_3$)$_2$(H$_2$O)$_2$(CN)$_2$]$^+$ (Figure 5.11) and *cis,cis,cis*-[Co(NH$_3$)$_2$(H$_2$O)$_2$(NO$_2$)$_2$]$^+$ [25]. It is noteworthy that in OC-6 a tris-heteroleptic coordination unit with monodentate ligands only can already be chiral, whereas in T-4 *four* different ligands are needed.

With the Bailar Tableau and the Mayper method for determining the chiral isomers, all possible isomers for OC-6 complexes with monodentate ligands can

Table 5.3. Bailar Tableau giving all 15 diastereomers for an OC-6 tris-heteroleptic complex $[Ma_2b_2c_2]^a$

	L	M	N
1	***ab***	ab	**ab**
	ca	cb	**cc**
	bc	ac	**ab**
2	ac	**ac**	ac
	ba	**bb**	bc
	bc	**ac**	ab
3	aa	**aa**	**aa**
	bc	**bb**	**bc**
	bc	**cc**	**cb**
4	ab	ab	ab
	bc	ba	bc
	ac	cc	ca
5	ac	ac	ac
	bc	ba	bb
	ab	cb	ca

ᵃFive diastereomers are indicated in **bold** and the chiral configuration in ***bold italic*** print.
2M≡5N; 1L≡1M≡2L≡2N≡4L≡4N≡5L≡5M; 3L≡3N; 1N≡4M.

easily be derived without using complicated mathematical formalisms. Table 5.4 gives the number and kind of isomers.

All complexes having C_s symmetry are *prochiral*. The pair of enantiotopic ligands are those which are related by the S_1 axis (the mirror plane); e.g. the complex $[Ma_4bc]$ has four homomorphic a ligands, numbered from 1 to 4 in Figure 5.12. They are all pair-wise heterotopic.

Of the six pairs, five (a_1/a_2, a_1/a_3, a_1/a_4, a_2/a_3, a_2/a_4) are diastereotopic, whereas a_3/a_4 is enantiotopic. In C_s-symmetric $[Ma_3b_2c]$ (Figure 5.13), the two b ligands are enantiotopic, whereas the a ligands form three pairs with a_1/a_2 being enantiotopic and a_1/a_3, a_2/a_3 diastereotopic.

$[Ma_2b_2cd]$ and $[Ma_2bcde]$ correspond to the prochiral SP-4 complexes, where two homomorphic, heterotopic 'ligands' are the lone pair phantom ligands. Two enantiotopic ligands in prochiral complexes can be designated in analogy with organic chemistry as pro-C or pro-A, since the C/A descriptors use the steering wheel reference system, as does the R/S nomenclature. For this purpose, the ligand under consideration is arbitrarily given a higher priority than its enantiotopic partner.

An example of a complex where the concepts of prostereoisomerism can be applied is the Ru^{II} complex depicted in Figure 5.14 [16]. The two homomorphic, enantiotopic chloride ligands are designated as pro-C and pro-A, respectively.

If $R' = H$, the overall symmetry (in the most symmetric conformation) is C_s. The two chloride ligands are enantiotopic, yielding a pair of enantiomers (C or A, respectively) upon substitution (hydrolysis) of one of them. The methyl groups of the dimethyl sulfoxide ligands *trans* to the chloride ligands are pair-wise

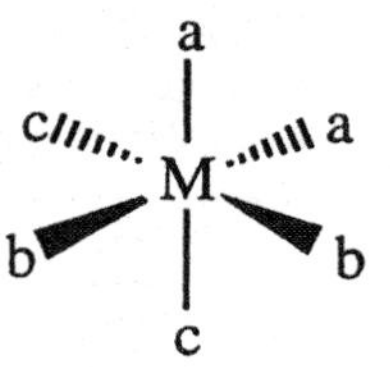

cis,cis,cis-$[Ma_2b_2c_2]$
OC-6-32-C-$[Ma_2b_2c_2]$
Highest possible symmetry: C_1

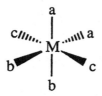

cis,cis,trans-$[Ma_2b_2c_2]$
OC-6-22-$[Ma_2b_2c_2]$
Highest possible symmetry: C_{2v}

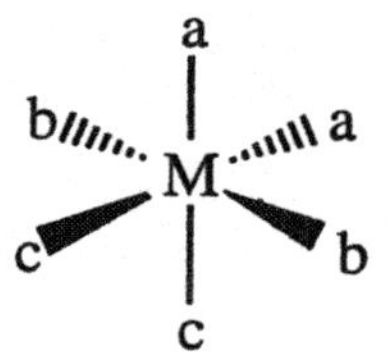

cis,trans,cis-$[Ma_2b_2c_2]$
OC-6-33-$[Ma_2b_2c_2]$
Highest possible symmetry: C_{2v}

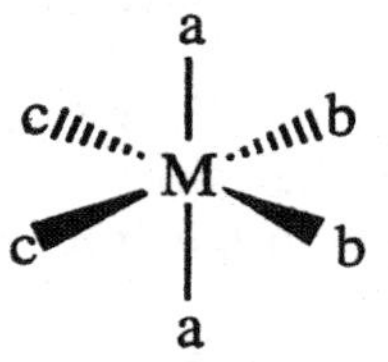

trans,cis,cis-$[Ma_2b_2c_2]$
OC-6-13-$[Ma_2b_2c_2]$
Highest possible symmetry: C_{2v}

trans,trans,trans-$[Ma_2b_2c_2]$
OC-6-12-$[Ma_2b_2c_2]$
Highest possible symmetry: D_{2h}

Figure 5.10
Representation of the five diastereomers of an OC-6 $[Ma_2b_2c_2]$ complex, with the corresponding configuration indices. For OC-6-32, the C absolute configuration is given

enantiotopic (Me(8)/Me(12) and Me(9)/Me(13)) and diastereotopic (Me(8)/Me(9) and Me(12)/(13)), respectively. The methyl groups of the dimethyl sulfoxide *trans* to the amine are enantiotopic. The NMR spectrum of this complex mirrors this distribution of methyl groups [16].

If a chiral amine (for a more general discussion of ligands containing chiral ligands, see Section 5.5) is coordinated instead of ammonia [e.g. R' = (S)- or (R)-CH(CH₃)-Ph], the C_s symmetry of the complex is broken by the stereogenic center in the ligand. All homomorphic ligands are thus diastereotopic in this complex.

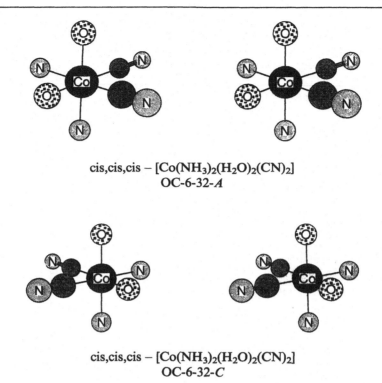

cis,cis,cis – [Co(NH$_3$)$_2$(H$_2$O)$_2$(CN)$_2$]
OC-6-32-A

cis,cis,cis – [Co(NH$_3$)$_2$(H$_2$O)$_2$(CN)$_2$]
OC-6-32-C

Figure. 5.11. The pair of enantiomeric *cis,cis,cis* – [Co(NH$_3$)$_2$(H$_2$O)$_2$(CN)$_2$]$^+$ complexes

The two chlorides, for example, yield two diastereomers (with different rate of formation) upon hydrolysis. Also all methyl groups will be diastereotopic, and hence inequivalent in the NMR spectrum.

Statistical considerations. Diastereomers will have, of course, different thermodynamic stabilities, i.e. they correspond to minima of different depths on the energy surface of the atomic assembly. If equilibrium is established, the concentrations will be determined by the Gibbs free energy differences between the isomers. The two enantiomers of a chiral pair will have the same Gibbs free energy in a nonchiral or in a racemic environment. The Gibbs free energy differences of the diastereomers will be determined by enthalpic and by entropic factors. One contribution to the entropic factor is easily evaluated from stereochemical considerations. Take as an example [Ma$_4$b$_2$]. It has two isomers, *trans* with D_{4h} symmetry and *cis* with C_{2v} symmetry. In an octahedron, there are three possibilities to put two ligands in a *trans* position, but there are twelve possibilities to put them in a *cis* position, since there are twelve edges. The statistical ratio is therefore *trans* : *cis* = 3 : 12. Equilibrated solutions therefore contain four times as much *cis* complexes as *trans* complexes if the enthalpic contributions are negligible. A similar consideration yields a ratio for *fac* : *mer* = 8 : 12 in [Ma$_3$b$_3$] complexes. Table 5.5 contains the statistical weights of the various isomers in the case of the heteroleptic OC-6 complexes with monodentate ligands.

Table 5.4. Number and kind of isomers and their highest possible symmetries for all OC-6 complexes with monodentate ligands

Complex type	Total No of isomers	Enantiomeric pairs	Highest possible symmetry
Homoleptic			
$[Ma_6]$	1	0	O_h
Bis-heteroleptic			
$[Ma_5b]$	1	0	C_{4v}
$[Ma_4b_2]$	2	0	D_{4h}, C_{2v}
$[Ma_3b_3]$	2	0	C_{2v}, C_{3v}
Tris-heteroleptic			
$[Ma_4bc]$	2	0	C_{4v}, C_s
$[Ma_3b_2c]$	3	0	C_{2v}, C_s
$[Ma_2b_2c_2]$	6	1	C_1, C_{2v}, D_{2h}
Tetra-heteroleptic			
$[Ma_3bcd]$	5	1	C_1, C_s
$[Ma_2b_2cd]$	8	2	C_1, C_s, C_{2v}
Penta-heteroleptic			
$[Ma_2bcde]$	15	6	C_1, C_s
Hexa-heteroleptic			
$[Mabcdef]$	30	15	C_1

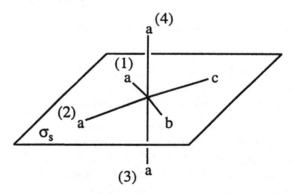

Figure 5.12
Example of a complex $[Ma_4bc]$ with four homomorphic (a_1, a_2, a_3, a_4), diastereotopic $(a_1/a_2, a_1/a_3, a_1/a_4, a_2/a_3, a_2/a_4)$, and enantiotopic (a_3/a_4) ligands in C_s symmetry

A case where nearly statistical ratios can be observed are the complexes $[PtBr_xCl_{6-x}]^{2-}$, where the concentrations of equilibrated solutions can easily be observed by ^{195}Pt NMR spectroscopy [26].

The application of the Bailar method for chelates will be discussed later.

TP-6. The VSEPR model makes the occurrence of $[Ml_6]$ coordination units displaying TP-6 fairly unlikely. Complexes with d^0 and d^{10} configurations are expected to have OC-6 CG following the VSEPR model. Complexes having d^n

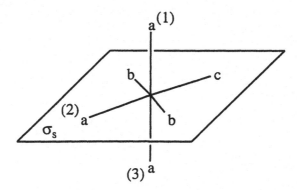

Figure 5.13
Example of a complex [Ma₃b₂c] with an enantiotopic pair of b ligands. The a ligands form one enantiotopic pair a_1/a_2, whereas a_1/a_3 and a_3/a_3 are diastereotopic

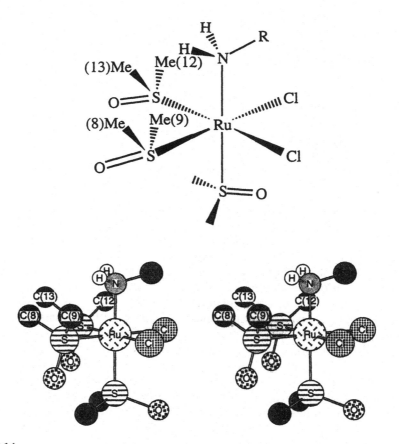

Figure 5.14
Prochiral *cis,fac*-[RuCl₂(DMSO)₃(NH₃)]. The two enaniomeric pairs of diastereotopic groups Me(8) and Me(12); Me(9) and Me(13) are indicated

Table 5.5. Statistical weights (absolute and relative) for the various isomers of homoleptic OC-6 complexes with monodentate ligands

Complex type	Isomer	Statistical weight	
		Absolute	Relative
Bis-heteroleptic			
$[Ma_4b_2]$	*cis* (C_{2v})	12	4
	trans (D_{4h})	3	1
$[Ma_3b_3]$	*fac* (C_{3v})	8	2
	mer (C_{2y})	12	3
Tris-heteroleptic			
$[Ma_4bc]$	*cis* (C_s)	24	4
	trans (C_{4v})	6	1
$[Ma_3b_2c]$	*fac* (C_s)	24	2
	mer (C_s)	24	2
	mer (C_{2v})	12	1
$[Ma_2b_2c_2]$	*cis, cis, cis* (C_1)	48	8
	cis, cis, trnas (C_{2v})	36	6
	trans, trans, trans (D_{2h})	6	1
Tetrakis-heteroleptic			
$[Ma_3bcd]$	*fac* (C_1)	48	2
	mer (C_s)	72	3
$[Ma_2b_2cd]$	*cis, cis* (C_1)	120	10
	trans, cis (C_s)	48	4
	trans, trans (C_{2y})	12	1
Pentakis-heteroleptic			
$[Ma_2bcde]$	*cis* (C_1)	288	4
	trans (C_s)	72	1
Hexakis-heteroleptic			
$[Mabcdef]$	All isomers	720	1

configurations that are not stable in OC-6 CG due to a Jahn–Teller instability, are not likely to distort towards TP-6. However, simple theories have their limitations. As was shown recently [27], d^0 [Ml_6] complexes can, under certain conditions, exhibit TP-6 CG owing to a so-called second-order Jahn–Teller effect. The conditions are: strong covalency of the M-l σ-bond and negligible π interaction. *Ab initio* calculations show that under such conditions TP-6 can have a lower energy than OC-6. Experimentally it was shown that [W(CH$_3$)$_6$] has TP-6 CG [19] (Figure 5.15), even in the gas phase, where effects from the surroundings cannot determine the CG.

Another example is [Li(TMEDA)$_2$]$_2$[Zr(CH$_3$)$_6$], which shows TP-6 in the solid [28] for the d^0 Zr^{IV} central metal. A simple-minded explanation of this phenomenon can be given in terms of strength of covalent bonding, using valence bond hybridization language: if covalent bonding with d-participation is important, TP-6 with hybridization d^{5}s and/or d^{3}sp^2 offers more bonding possibilities then OC-6 with d^{2}sp^3 hybridization, thereby stabilizing the former with respect to the latter CG.

The lower symmetry of the idealized TP-6 polyhedron, as compared with OC-6, yields a larger number of possible isomers in hypothetical heteroleptic complexes. The hexa-heteroleptic case gives a total of 118 stereoisomers, i.e. 59 pairs of

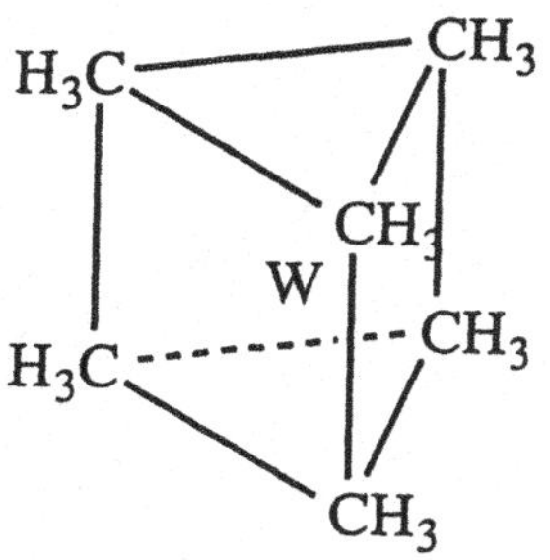

Figure 5.15
TP-6 [W(CH$_3$)$_6$], a rare case of a [Ma$_6$] complex which does not have OC-6, CG

enantiomers. The configuration index is a three digit-number. We shall not consider isomers of heteroleptic complexes for TP-6 and any higher CNs in detail, because stereochemically rigid, heteroleptic complexes with monodentate ligands are not known.

The relationship between OC-6 and TP-6 and possible transitions between these coordination geometries will be discussed in connection with the bidentate ligands.

5.2 Coordination Units with Achiral, Monodentate Ligands Only. Conformations of Monodentate Ligands

As was demonstrated in 1936 by Pitzer, an energy barrier of about 13kJmol^{-1} has to be assumed for the internal rotation of one CH$_3$ group with respect to the other in the ethane molecule. That observation led to the concept of *conformers*, and later to *conformational analysis*, which has become an important area of organic stereochemistry [29]. Mono- and polycentric ligands, l$_m$ and l$_p$, are often amenable to the same type of analysis, where internal rotation in the ligands themselves, or about the metal to ligand bond, has to be taken into account. Consider a simple case of an [M(l$_m$)$_6$] complex. As pointed out before, it will not often have exact O_h symmetry because the ligands do not comply with all the symmetry elements of O_h. This is the case, e.g., for l$_m$ = H$_2$O. VSEPR predicts a TPY-3 'coordination' for the ligating oxygen atom. Internal rotation about the M—O bond (Figure 5.16), corresponds to a change in conformation of the [M(H$_2$O)$_6$] complex. It may be conjectured [30] that a 'high' symmetry conformer (Figure 5.17) in which ligand–ligand interaction is optimized through internal H-bonding, corresponds to a local or even global energy minimum. The symmetry of this conformer is C_i, and the complex is therefore not chiral.

A thorough analysis of the [Mg(H$_2$O)$_6$]$^{2+}$ ion by experimental and theoretical methods indicates that there is no likelihood of intramolecular hydrogen bonding in the coordination polyhedron of hydrated Mg^{2+} [31]. The model depicted in Figure 5.17 indicates that such an arrangement is highly unlikely for any hydrated metal ion, since the geometry of the four electron pairs surrounding one oxygen ligand atom would deviate drastically from the T-4 arrangement required by the VSEPR theory.

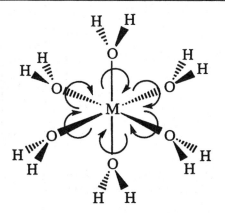

Figure 5.16
Conformational changes in an OC-6 [M(H$_2$O)$_6$] complex through rotations of the OH$_2$ ligands about the M—O bond directions

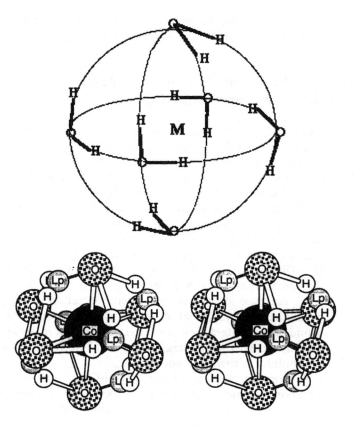

Figure 5.17
Arrangement of symmetry C_i, of the OH$_2$ ligands in an [M(H$_2$O)$_6$] complex, if intramolecular hydrogen bonding were important. (a) Schematic representation; (b) stereo pair (lp = lone pair)

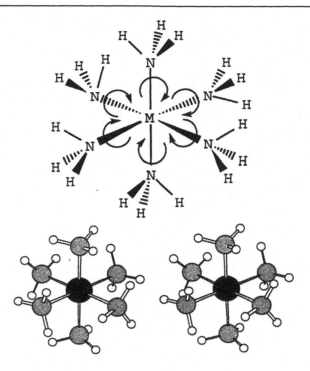

Figure 5.18
Conformational changes in an OC-6 [M(NH$_3$)$_6$] complex through rotations of the ligands about the M—N bond directions

A similar complex [M(NH$_3$)$_6$] is expected to behave somewhat differently, because the coordinating nitrogen is now a T-4 center. The most symmetric arrangement of the six ammonia ligands is C_s (Figure 5.18).

A conformational analysis of this kind is very difficult, both experimentally and theoretically. Theoretically, because such complexes are strongly solvated and calculations of isolated species are therefore not relevant. Experimentally, because solution methods are not easily, if at all, available for the investigation of the conformation of the hydration or solvation sphere in general. Of the various diffraction methods in the solid, only neutron diffraction might yield some answer to positions of hydrogen atoms in the above-mentioned cases. Not many confirmed investigations are therefore available, and the stereochemistry concerning protons bonded to ligating atoms, such as water and ammonia is essentially a 'white region' in the map of coordination compounds, with some exceptions such as the [Mg(H$_2$O)$_6$]$^{2+}$ mentioned above.

Conformational analysis of organic ligands has been a subject of increasing interest in coordination chemistry since the classical work of Corey and Bailar [32] on chelate complexes, and it has since become a topic subject [33,34]. Conformations of chelate complexes will be discussed in Section 5.3.2. In some cases, conformations have been discussed also for non-chelating ligands. A thorough conformational analysis is given, e.g., for a series of complexes

$[(C_5H_5)Fe(CO)(PPh_6)(R')]$ with several carbon-bonded R ligands [6]. The reader is referred to this review for further details.

5.3 Coordination Units Comprising Achiral, Bidentate Ligands

5.3.1 Planar Chelate Rings

Complexes comprising ligands with a denticity >1 are topologically[†] different from the coordination units with monodentate ligands. If in the latter a center (metal) to ligand bond is broken, the molecule is divided into two independent parts. In chelate complexes (Figure 5.19), the breaking of a single metal to ligand bond preserves the molecular unity.

There are many consequences of this, especially a generally enhanced thermodynamic stability of the complexes, which is often called the chelate effect [35,36]. There are also many stereochemical consequences of the presence of chelate ligands in a coordination unit. In the present chapter, we discuss the most simple type of chelate ligands, bidentate structures (A=A) or (A=B), i.e. ligands, where all atoms of the chelate ring lie in a plane yielding a planar chelate ring. The other class of bidentate ligands, (A≈A) or (A≈B), where the chelate ring is

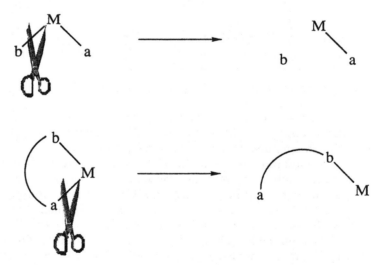

Figure 5.19
The topological difference between chelate and non-chelate complexes

[†]Geometric topology is a branch of mathematics, which deals with those properties of geometric objects for which their shape and their size are unimportant. No use is made in our discussions of any formal mathematical theory of topology. We distinguish here for practical reasons topographical properties of molecules such as number and kind of isomers, and *topological* properties which concern uniquely connectivity properties. The terms *topology* and *topological* have been and still are often used in chemistry in a somewhat slack way. The term topological isomer should be reserved for those cases where a molecule cannot be transformed into an isomer without breaking bonds.

puckered, behave similarly in many respects, but in others in a distinctly different manner. The present discussion will therefore cover many aspects of the stereochemistry of non-planar chelating ligands also. Those properties, which are specific for puckered chelate rings, will be discussed in a chapter dealing with ligand conformations (Section 5.3.2).

Planar chelate rings can, in principle, have any number of atomic centers $\geqslant 3$, including the coordination center. We do not consider here the case of three membered rings, i.e. complexes where the ligating atoms are directly bound to each other (such as O_2, H_2N—NH_2, H_2C=CH_2 etc.), because it can always be argued as to whether the bonding from the metal center to the ligand is directed to the ligand atoms or to the bond between these atoms. Besides that, such ligands do not give rise to many complexes differing in their topographical stereochemistry. Metrical stereochemical properties of such ligands have been discussed in the literature, for instance the important question of the geometry of the $M \cdots O_2$ interaction.

The ligands that are of interest here give mainly four-, five-, or six-membered chelate rings. Some examples of such ligands are (5.7) (see also Appendix III).

For steric reasons, only *cis* positions can be occupied in SP-4 and in OC-6

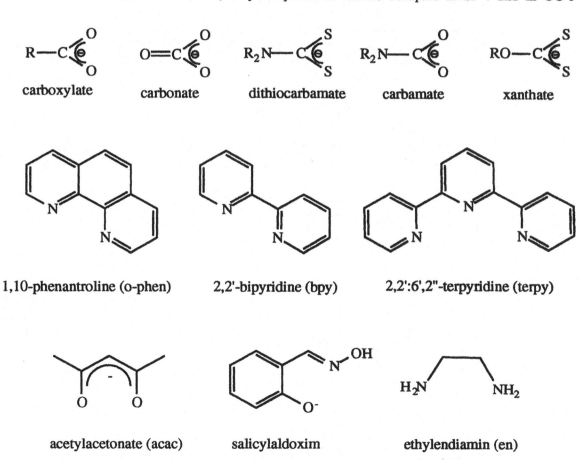

carboxylate   carbonate   dithiocarbamate   carbamate   xanthate

1,10-phenantroline (o-phen)   2,2'-bipyridine (bpy)   2,2':6',2"-terpyridine (terpy)

acetylacetonate (acac)   salicylaldoxim   ethylendiamin (en)

(5.7)

coordination geometries with the ligands mentioned above [as with most (A≈A) ligands]. Ligands have been synthesized which can span *trans* positions. These cases will be discussed separately, and in the forthcoming discussion we assume tacitly that the bidentate ligands can occupy only *cis* positions.

Coordination Number 4

The restriction mentioned in the title of this section is not very important in reality, since most of the stereochemical behavior of such complexes is similar with ligands forming planar and those forming puckered rings. Therefore, the general symbols (A^A) and (A^B) will be used.

T-4 (Figure 5.20). This case is analogous to the well known tetrahedral carbon center. Again, the same comparisons concerning inertness with respect to substitution reactions can be made between a C- and an M-center. Coordination units [M(A^A)a$_2$] and [M(A^A)ab] have symmetries C_{2v} and C_s, respectively, and they cannot give rise to stereoisomers. The same is true for [M(A^A)$_2$] with D_{2d} symmetry. On the other hand, [M(A^B)ab] and [M(A^B)$_2$] have C_1 and C_2 symmetries, respectively, and they are therefore chiral. The former creates no problems with the usual CIP rules for the ligands, because every ligand atom has a different priority assigned to it. The complex [M(A^B)$_2$], on the other hand, needs a supplement to the basic CIP rules. However, its absolute configuration can easily be assigned within the oriented line reference system.

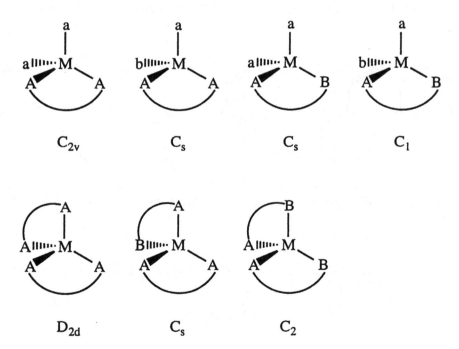

Figure 5.20
All possible arrangements of T-4 complexes with one or two bidentate chelate complexes. No diastereomers occur. [M(A^B)ab] and [M(A^B)$_2$] are chiral

In organic chemistry, there are many examples known of chiral, so-called spiro-compounds. These are molecules of the type $[C(A\hat{\ }B)_2]$. In most cases spiro compounds do not racemize. In coordination chemistry, complexes with non-symmetrical chelates coordinating tetrahedrally to a metal center are also known. Owing to rapid ligand-exchange reactions or/and polytopal isomerism, EPCs of the latter are not often obtained. It has been shown, for example, that the complex formed between Ni^{II} and N-alkylsalicylaldimine chelates shows T-4 and SP-4 coordination geometry in the solid state and in solution. Racemization (Figure 5.21) in T-4 would therefore be extremely rapid (see also Figure 5.2).

Some T-4 complexes have been resolved into the enantiomers, notably $[Be(benzoylpyruvate)_2]^{2-}$ [37] and $[Zn(8\text{-quinolato-5-sulfonic acid})_2]$ [38]. The chirality of these complexes can obviously be described by the oriented line system. The D_{2d} symmetry of a T-4 $[M(A\hat{\ }A)_2]$ complex is reduced to C_2 in T-4 $[M(A\hat{\ }B)_2]$, but the two chelate rings remain orthogonal. The designation for the two enantiomers is shown in Figure 5.22. Note that the chirality descriptor does not depend upon the choice of the orientation of the line.

Ligands with special topological properties, which still comply however with the definition of a bidentate chelate ligand, have found recent interests in so-called catenands and in similar molecular constructs. Those ligands and their complexes will be discussed later.

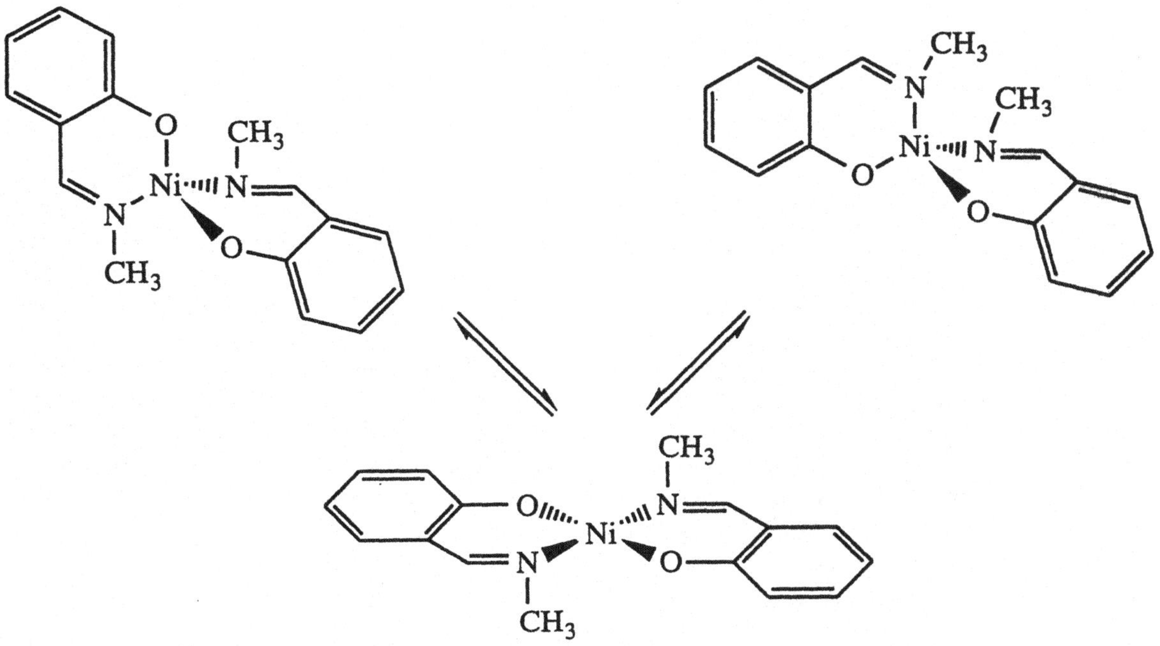

Figure 5.21
Racemization of a T-4 $[M(A\hat{\ }B)_2]$ complex through its polytopal isomer SP-4

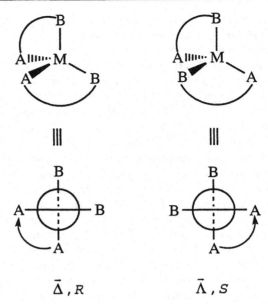

Figure 5.22
Chirality descriptors for a T-4 [M(A^B)₂] complex

$$[M(A=A)ab] \qquad\qquad [M(A=B)ab]\ [M(A=B)a_2]$$
$$[M(A=A)a_2]$$

$$[M(A=A)_2]$$
$$[M(A=A)(B=B)]$$
$$[M(A=B)_2]$$

(5.8)

SP-4. There are many examples of complexes in this geometry for the various constitutions (5.8).

With the elements of Group 8 a great number of different isomers have been synthesized, especially with Pt^{II}, because these complexes are often stereochemically rigid, and they follow in their reactivity certain relatively simple rules which will be discussed later. The inherent coplanarity of the central atom and the ligand atoms in SP-4 excludes the possibility of chiral compounds if the chelates are planar. We shall have to consider carefully, however, the consequences of deviations from strict planarity for the occurrence of chiral complexes. Such cases will be discussed in Section 5.3.2.

Coordination Number 5

Chelate ligands can make complexes of this coordination number much more rigid, as compared with complexes with exclusively monodentate ligands. In the case of $[M(cp)(CO)_2(A^B)]$, even enantiomerically pure or at least enantiomerically

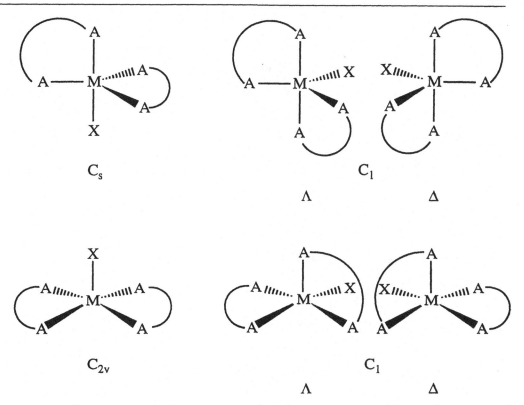

Figure 5.23
Possible configurations of 5-coordinated complexes [M(A=A)₂X] in TP-5 and in SPY-5

enriched compounds were obtained in TP-5 CG [39]. Complexes in either TB-5 or in SP-5 may give rise to one non-chiral and one chiral diastereomer. For symmetrical ligands (A=A), the possible configurations are given in Figure 5.23.

In each of the two coordination geometries one asymmetric C_1 arrangement exists, which gives rise to pairs of enantiomers. Chirality descriptors are therefore needed to characterize the stereoisomers completely. It can be anticipated, however, that a separation of enantiomers of complexes [M(A^A)₂X] will hardly be possible, because the mechanism of pseudorotation is still feasible and it will make racemization a fast process. We therefore postpone the introduction of the appropriate chirality descriptors to the paragraph about OC-6 complexes, where they are used frequently.

Kepert [40] analyzed some (Ru, Sb, Sn, Fe) [M(A^A)(l)₃] and a few (Os, V, Mo, Re, Fe, Co, Ni, Cu, Zn, Ru, Ir, Cd) [M(A^A)₂(l)] complexes in the frame of the repulsion model. The reader is referred to this discussion for further details.

Coordination Number 6

OC-6. There are coordination units comprising one, two or three bidentate ligands. Various types of constitutions are possible (5.9).

mono-chelates

[M(A^A)abcd]   [M(A^B)abcd]

[M(A^A)$a_2$bc]   [M(A^B)$a_2$bc]

[M(A^A)a_2b_2]   [M(A^B)a_2b_2]

[M(A^A)a_3b]   [M(A^B)a_3b]

[M(A^A)a_4]   [M(A^B)a_4]

bis-chelates

[M(A^A)$_2$ab]   [M(A^B)$_2$ab]

[M(A^A)$_2a_2$]   [M(A^B)$_2a_2$]

[M(A^A)(B^C)ab]   [M(A^A)(B^C)a_2]

[M(A^B)(C^D)ab]   [M(A^B)(C^D)a_2]

tris-chelates

[M(A^A)$_3$]   [M(A^B)$_3$]

[M(A^A)$_2$(B^B)]   [M(A^A)$_2$(B^C)]   [M(A^A)(B^C)(D^E)]

[M(A^A)(B^B)(C^C)]   [M(A^A)(B^C)$_2$]   [M(A^B)(C^D)(E^F)]

$$(5.9)$$

Every constitution yields several possibilities for stereoisomers, and some investigations have become classics [41–43] in coordination chemistry, e.g. the preparation of the diastereomers of $[CoCl_2(NH_3)_2(en)]^+$ by Bailar and Peppard. Before discussing some of their properties, an extension of the nomenclature concerning stereochemical descriptors is introduced. Consider a coordination unit comprising at least two bidentate ligands (symmetrical (A=A) for the sake of simplicity). There are three and only three different configurations of these two chelates, if coordinated to an octahedron (Figure 5.24).

The problem of finding all isomers of a given constitution with complexes of bidentate ligands can again be solved using the Bailar Tableau. Assuming that a chelate ligand cannot span *trans* positions in OC-6 complexes, ligand atoms of the same ligand therefore cannot form a pair in the list. If two or three identical ligands are present, the ligands have to be labeled. As an example, the 'word processor' method is applied for the complex [M(A^A)$_3$] (Table 5.6):

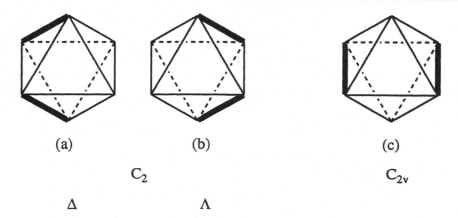

(a)   (b)   (c)

C_2 C_{2v}

Δ Λ

Figure 5.24
Chiral *cis* (a and b) and achiral *trans* (c) configurations in an OC-6 [M(A$^\wedge$A)$_2$X$_2$] coordination unit; (c) obviously corresponds to a nonchiral configuration (symmetry C_{2v}), since the two ligands are coplanar (see, however, the remark made concerning SP-4 complexes with ligands deviating from planarity!). On the other hand, (a) and (b) form a pair of enantiomers, with symmetry C_2. Here the Δ/Λ chirality descriptors (Section 4.4) are clearly appropriate and they are used throughout for OC-6 complexes having at least two chelate rings

Table 5.6. Bailar Tableau for an OC-6 [M(A$^\wedge$A)$_3$] complex

	L	M	N
1	AA	AA	AA
	A'A'	A'A''	A'A''
	A''A''	A'A''	A'A''
2	AA'	AA'	AA'
	AA'	AA''	AA''
	A''A''	A'A''	A'A''
3	AA'	AA'	AA'
	AA'	AA''	AA''
	A''A''	A'A''	A'A''
4	AA''	AA''	AA''
	AA'	AA'	AA''
	A'A''	A'A''	A'A'
5	AA''	AA''	AA''
	AA'	AA'	AA''
	A'A''	A'A''	A'A'

<Replace a by A; b by A; c by A'; d by A'; e by A''; f by A''>; 1L, 1M, 1N, 2L, 3L, 4N, 5N contain all *trans* pairs of one chelate ligand, and they are therefore eliminated. The other eight entries are all equivalent. Therefore, the only diastereomer of this composition is a chiral structure, as of course is obvious in this simple case. Clearly, the Δ/Λ descriptors can be applied for the designation of the chirality of [M(A=A)$_3$] (Figure 5.25). A chiral molecular entity can have two

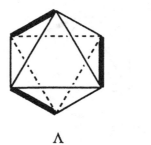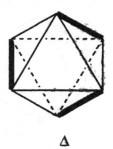

Λ $\qquad\qquad\qquad\qquad\qquad\qquad$ Δ

Figure 5.25
OC-6 [M(A=A)$_3$] in its two enantiomeric forms

and only two enantiomeric isomers. One pair of chirality descriptors (e.g., Δ/Λ) suffices therefore for their characterization. We could indicate all pairwise orientations of each pair of skew ligands, which would be Δ,Δ,Δ and Λ,Λ,Λ for [M(A^A)$_3$], respectively. This pair-wise designation will be of some importance for the multidentate ligands. In the projection along the threefold axis, the helical arrangement can be easily associated with a left- and a right-handed screw, respectively.

The number and types of isomers in OC-6 complexes with bidentate ligands, combined with monodentate ligands, is given in Table 5.7.

There is a plethora of chiral OC-6 [M(A=A)$_3$] complexes, and in many cases the enantiomers have been purified, most often by formation of diastereomeric salts with enantiomerically pure counter ions. Tris-acetylacetonate complexes of metals in +III oxidation state are neutral, however, and the method of salt formation cannot be applied. In the case of [Cr^{III}(CF$_3$COCHCOCF$_3$)$_3$], separation into enantiomers on chiral gas chromatographic columns has been achieved [44]. An interesting case of tris(bidentate) OC-6 complexes with purely inorganic ligands is represented by the ammonium platinum sulfides (NH$_4$)$_2$[Pt(S$_5$)$_3$](H$_2$O)$_2$ and (NH$_4$)$_2$[Pt(S$_6$)$_2$(S$_5$)](H$_2$O)$_2$ [45–47] (Figure 5.26).

These compounds crystallize from aqueous solution, the S$_{15}$ complex at higher pH and the S$_{17}$ species at lower pH values. The S$_{15}$ complex can be separated into the enantiomers using [Ru(bpy)$_3$]$^{2+}$, whereas the S$_{17}$ complex spontaneously crystallizes as an enantiomerically pure compound if the conditions are controlled carefully around pH 9.2! Apparently, the formation of the one enantiomer as the only product is a kinetically controlled reaction, whereby chiral seeds induce the growth of crystals of one enantiomer. A rapid solution process, probably involving a Pt^{II} species, replenishes the concentration of this enantiomer efficiently, keeping the solution nearly racemic during the crystallization process.

Three ligands (A^B) give, of course, two pairs of enantiomers that can be designated as *fac* and *mer*, respectively. The former has C_3 and the latter C_1 symmetry. In diamagnetic complexes with organic ligands, such as in [Co^{III}(glycinato)$_3$] the two diastereomers can easily be distinguished by ^{13}C NMR spectroscopy, where the *fac* isomers shows three and the *mer* isomer six signals [48].

TP-6.   TP-6 is simpler than OC-6 in the respect, in that the *cis* positions define parallel lines and the complexes are therefore in general not chiral. In fact, TP-6 is

Table 5.7. Number and types of complexes of OC-6 complexes with bidentate and monodentate ligands simultaneously present

Type of complex	Total No. of isomers	No. of chiral pairs	Symmetries
[M(A═A)abcd]	12	6	C_1
[M(A═B)abcd]	24	12	C_1
[M(A═A)$_2$ab]	3	1	C_{2v}, C_1
[M(A═B)$_2$ab]	11	5	C_s, C_2, C_1
[M(A═A)a$_2$bc]	7	3	C_s, C_1
[M(A═B)a$_2$bc]	12	5	C_s, C_1
[M(A═A)$_2$a$_2$]	3	1	D_{2h}, C_2
[M(A═B)$_2$a$_2$]	8	3	C_{2h}, C_{2v}, C_2, C_1
[M(A═A)a$_3$b]	2	0	C_s
[M(A═B)a$_3$b]	4	1	C_s, C_1
[M(A═A)(B═C)ab]	10	5	C_1
[M(A═A)(B═C)a$_2$]	5	2	C_s, C_1
[M(A═A)a$_4$]	1	0	C_{2v}
[M(A═B)a$_4$]	1	0	C_s
[M(A═B)(C═D)ab]	20	10	C_1
[M(A═B)(C═D)a$_2$]	10	4	C_s, C_1
[M(A═A)$_3$]	2	1	D_3
[M(A═B)$_3$]	4	2	C_3, C_1
[M(A═A)$_2$(B═B)]	2	1	C_2
[M(A═A)$_2$(B═C)]	2	1	C_1
[M(A═A)(B═C)(D═E)]	8	4	C_1
[M(A═A)(B═B)(C═D)]	4	2	C_1
[M(A═A)(B═B)(C═C)]	2	1	C_1
[M(A═A)(B═C)$_2$]	6	3	C_1
[M(A═B)(C═D)(E═F)]	16	8	C_1

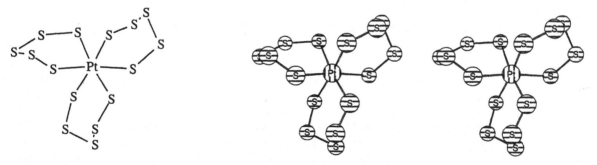

Figure 5.26
A stereopair of the Δ-form of [Pt(S$_5$)$_3$]$^{2-}$

a possible non-chiral transition state for a racemization process in chiral OC-6 complexes. This will be discussed in more detail in Chapter 7, Section 7.1.

Relationship between OC-6 and TP-6. For a complex [M(A═A)$_3$], the deformation from OC-6 towards TP-6 can be given in terms of one angular parameter (Figure 5.27) and a value for the 'normalized bite' angle.

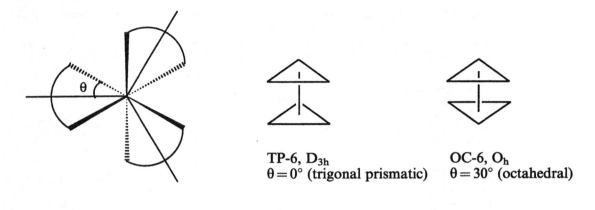

TP-6, D_{3h}
$\theta = 0°$ (trigonal prismatic)

OC-6, O_h
$\theta = 30°$ (octahedral)

$$b = 2 \sin \left(\frac{i\hat{M}j}{2} \right)$$

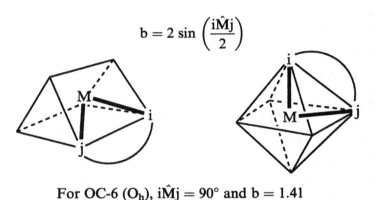

For OC-6 (O_h), $i\hat{M}j = 90°$ and $b = 1.41$

Figure 5.27
θ (twist angle) and normalized b (bite) for complexes ranging from OC-6 to TP-6

Kepert [40] analyzed 158 complexes in the frame of the repulsion model to determine the relationship between the 'normalized bite' and the angle θ. There is good agreement for the relationship between these two parameters as compared with the predictions of the repulsion model, with some exceptions. These are [49] all tris(bidentate) chelate complexes with dithio ligands.

Complexes with bidentate ligands showing coordination numbers larger than six have also been discussed by Kepert [40].

5.3.2   Bidentate Chelate Ligands Deviating from Planarity

Only some ligands, e.g. 2,2'-bipyridine, 1,10-phenanthroline, and other ligands with delocalized p-bonding in the chelate ring, can be strictly planar. However, even π-bonded molecules may deviate appreciably from planarity in metal complexes. Other ligands, those with aliphatic connections between the ligating atoms, are inherently non-planar. One can distinguish three different types of non-planarity in such complexes: (a) inherently planar ligands, which become non-planar through ligand–ligand interactions; (b) bidentate chelates with non-rigid, non-planar geometries, due to conformational properties of the ligand; and (c) ligands with 'rigid' non-planar geometries due to intra-ligand interactions (ring formation, atrop-isomerism). Typical examples for these three cases are given in Figure 5.28.

In this section, we will discuss only the cases (a) and (b), since (c) involves mostly chiral ligands, or at least ligands with stereogenic centers. These cases will be discussed in Section 5.5.

Ligand non-planarity due to ligand–ligand interactions. Ligands that are inherently planar in the non-coordinated form, such as 1,10-phen, or which can be brought into a planar configuration by changing dihedral angles and not bond angles, such as bpy (by rotation about the C—C bond, which connects the two pyridine rings), are often considered to be planar in the coordination unit. This is often the case to a good approximation. Under special circumstances, however, deviations from planarity can become appreciable. This has been observed, e.g., in SP-4 complexes, with two 'planar' aromatic ligands which interact so strongly that coplanarity of all atoms is no longer possible. Homoleptic bis-cyclometalated (Figures 5.28(a), and 5.30) complexes of Pt^{II} have been shown to have an extremely strong tendency to have SP-4 *cis* configuration [50]. The highest symmetry possible for such a complex is C_{2v}, if all atoms lie in a plane. If R is a molecular fragment (Figure 5.29) of the ligand, extending into space in such a manner that it interacts strongly with R from the other ligand, a distortion will occur.

The distortion can only lower the symmetry, and the distorted molecule must therefore belong to a subgroup of C_{2v}, i.e. C_2, C_s, or C_1. Distortion towards C_s would mean a simultaneous bending of the two R groups downwards (or upwards), which would not relieve the strain in the molecule. Distortion to C_2, on the other hand, corresponds to an upward and downward bending of the R groups from the molecular plane, respectively. This distortion can relieve the molecular strain. Seen from the direction indicated by the arrow in Figure 5.29, the ligands are arranged as depicted in Figure 5.30.

This is one of the simplest cases of a helical chiral molecule, and it represents a two-bladed propeller whose chirality can immediately be assigned using the Δ, Λ descriptors, introduced in Chapter 4, Section 4.4.

A case of such a complex is [Pt(diphpy)$_2$], which was the first chiral complex of this kind reported [51]. It does not represent not an isolated case, however, and several other similar complexes have been prepared (A. von Zelewsky *et al.*, unpublished work).

Ligand conformations, five-membered chelate rings. As pointed out for the first time by Corey and Bailar [32], some chelate rings such as 1,2-diaminoethane (en) cannot be planar without having grossly distorted bond angles and bond lengths. Typical structural data are given in Figure 5.31, and they correspond to an experimental (X-ray) determination in the complex [Co(en)$_3$][Ni(CN)$_5$], mentioned also in connection with CN 5 in Ref. 18 (Figure 5.31). Numerous other experimental determinations of conformations of the en-ring have been reported.

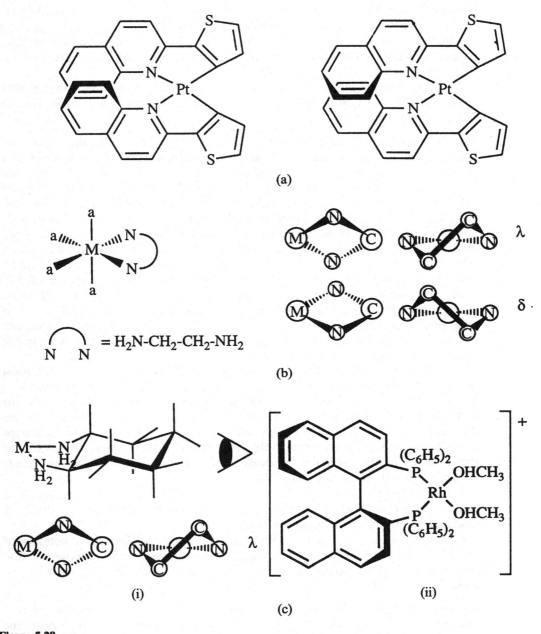

Figure 5.28
Three cases of nonplanar bidentate ligands in complexes. (a) [Pt(thq)$_2$], where H-thq = (2'-thienyl)-2-quinoline; (b) en = 1,2-diaminoethane (two conformations are possible, here [M(en)a$_4$] as an example); (c) [Rh(Binap)L$_2$]$^+$, where Binap = 2,2'-bis(diphenylphosphino)-1,1'-binaphthyl. For (S)-BINAP, the seven-membered chelate ring is fixed to the δ-skew-boat conformation.

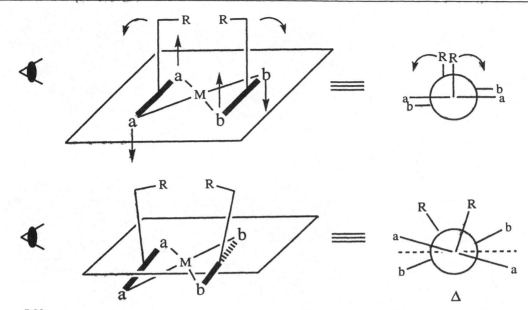

Figure 5.29
Schematic drawing of a SP-4 complex with interacting ligands; examples are depicted in Figure 5.30

The shape and the size of the cobalt-ethylenediamine ring (Figure 5.31) are as follows:

Co—N	1.987 ± 0.004Å	N—Co—N	$85.4 \pm 0.3°$
N—C	1.497 ± 0.010Å	Co—N—C	$108.4 \pm 0.5°$
C—C	1.510 ± 0.010Å	N—C—C	$105.8 \pm 0.7°$
		N—C—C—N	$55.0°$

The dihedral angle N—C—C—N is a measure of the deviation of the chelate ring from planarity. The two possibilities for the conformations are depicted in Figure 5.32.

It is obvious that the chelate ring is a helically chiral moiety with C_2 symmetry and the two conformers correspond to an enantiomeric pair. The descriptors for helical chirality are clearly appropriate. These conformational helical chiralities are designated by δ and λ, respectively (Figure 5.32).

A chiral conformation can also be obtained in the case of coordinated 2,2′-diaminobiphenyl as chelating ligand (Figure 5.33).

The most stable conformation of the free ligand is a non-coplanar orientation of the two phenyl rings with symmetry C_2. This symmetry is preserved upon complexation [52].

SP-4 and T-4 complexes [M(en)$_2$] will have each three stereoisomers due to ligand conformation, namely a racemic pair $\delta\delta/\lambda\lambda$ of C_2 symmetry and a non-chiral form $\delta\lambda$ of C_s symmetry. The energy differences between these conformations, and the energy barriers for changes of conformation, are small in these cases and consequently no isomers can be isolated. With 2,2′-diaminobiphenyl on the other hand, the activation energy for inversion of configuration is relatively high. The complex [Pt(2,2′-diaminobiphenyl)(en)]$^{2+}$ has been resolved into the enantiomeric

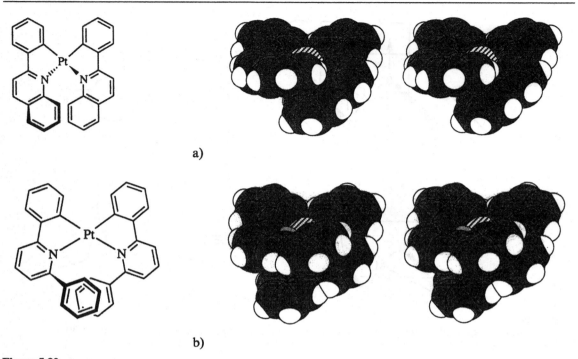

a)

b)

Figure 5.30
Two-bladed chiral SP-4 complexes. (a) [Pt(thq)$_2$]; (b) [Pt(phpy{ph-H})$_2$]

forms [52] and it does not racemize in solution at room temperature. No structural data seem to be available for this complex, and the dihedral angle for the two phenyl rings is therefore not known.

As already pointed out by Corey and Bailar in the first publication dealing with conformations of chelate rings, the really interesting case is the OC-6 [M(en)$_3$] coordination unit. Being inherently chiral through the helical arrangement of the bidentate ligands (Δ/Λ), this coordination unit has the following diastereomeric conformations, always appearing in enantiomeric pairs (Figure 5.34): $\Delta(\delta\delta\delta)/\Lambda(\lambda\lambda\lambda)$; $\Delta(\delta\delta\lambda)/\Lambda(\lambda\lambda\delta)$; $\Delta(\delta\lambda\lambda)/\Lambda(\lambda\delta\delta)$; $\Delta(\lambda\lambda\lambda)/\Lambda(\delta\delta\delta)$. The symmetry is D_3 for the homoconformational cases, but it is reduced to C_2 in the heteroconformational complexes.

As Figures 5.34(a–d) show, the chelate rings have a different projection if looked along the *threefold* axis of (a) and (d), or along the *pseudo-threefold* axis in the C_2-symmetrical cases (b) and (c). In (a), the C—C bond directions are nearly parallel to the threefold axis, making the second carbon atoms to appear hidden behind the top ones. In (d), the C—C bonds in the ligands are clearly visible, since their angles with the threefold axis is in the range of 30°. In the mixed cases both projections appear in the same complex. Owing to this property, the designations ob (oblique) and lel (parallel) for ligand conformations of this type has been introduced [32]. The use of these labels is sometimes very practical. It has to be pointed out that ob and lel do not, in general, correspond directly to the chirality descriptors δ and λ. The following relationships between these two different ways of designations are easily established:

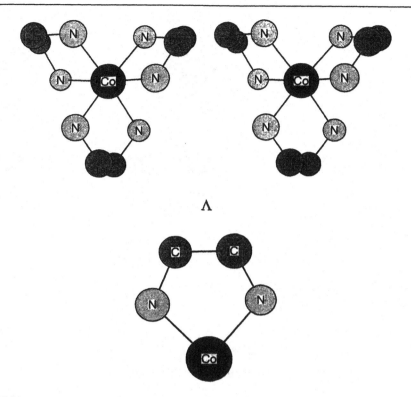

Figure 5.31
Λ-$[Co(en)_3]^{3+}$ and one of its chelate rings. For bond lengths and bond angles see p. 109

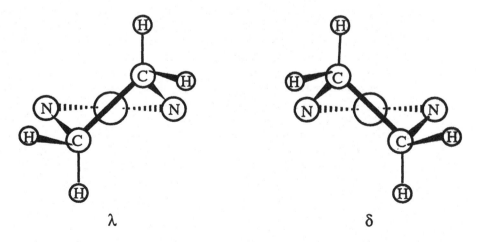

Figure 5.32
Two enantiomeric conformations of a five-membered chelate ring with T-4 carbon atoms in the chelate ring

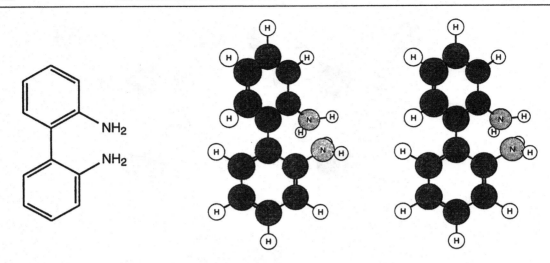

Figure 5.33
Stereo pair of a chelate ring with 2,2'-diaminobiphenyl ligand

$$\Lambda(\delta\delta\delta) \leftrightarrow \Lambda(\text{lel,lel,lel}) \text{ and } \Lambda(\lambda\lambda\lambda) \leftrightarrow \Lambda(\text{ob,ob,ob})$$
$$\Delta(\lambda\lambda\lambda) \leftrightarrow \Delta(\text{lel,lel,lel}) \text{ and } \Delta(\delta\delta\delta) \leftrightarrow \Lambda(\text{ob,ob,ob})$$

Lel and ob are furthermore only defined when there are at least two bidentate ligands coordinated to one metal, so that the two remaining positions are *cis*. The projection axis is the true or pseudo-C_3, .which relates the three *cis* positions in the octahedral unit. λ and δ, on the other hand are defined for each single chelate ring.

The four diastereomers of Figure 5.34 will in general have different chemical potentials in the thermodynamic sense. They will represent local minima on the energy hypersurface. However, it is not trivial to predict which conformation corresponds to the global minimum. The mixed conformations, $\delta\lambda\lambda$ and $\delta\delta\lambda$, have a higher statistical weight as compared with either of the pure conformations $\delta\delta\delta$ and $\lambda\lambda\lambda$, and they are therefore entropically preferred. In any case, the energy differences are relatively small and the energy barriers for inter conversion of the ligand conformations are small, making those processes rapid. Conformations of five-membered chelate rings have been determined in a great number of cases by X-ray diffraction [53–55], and conformational analyses have been carried out by NMR spectroscopy [33] and force field calculations [34]. Proton NMR spectroscopy is particularly suited for such investigations, since rates of transformation of ligand conformations are often in the range of observability by NMR methods.

Strain energy minimization studies of the ring inversion (which is a process with $\Delta G° = 0$ in a non-chiral, or in a racemic environment)

$$[\text{Co}(\delta\text{-en})(\text{NH}_3)_4]^{3+} \rightarrow [\text{Co}(\lambda\text{-en})(\text{NH}_3)_4]^{3+}$$

yielded 15.7kJmol^{-1} as energy barrier for inversion [56]. Similar calculations for N,N,N',N'-tetramethylethane-1,2-diamine (tmen) show a significant increase to 24.6kJmol^{-1}, whereas replacing the four ammonia ligands by two en chelates increases the inversion barrier just to 17.1kJmol^{-1} per chelate ring in [Co(en)$_3$]$^{3+}$.

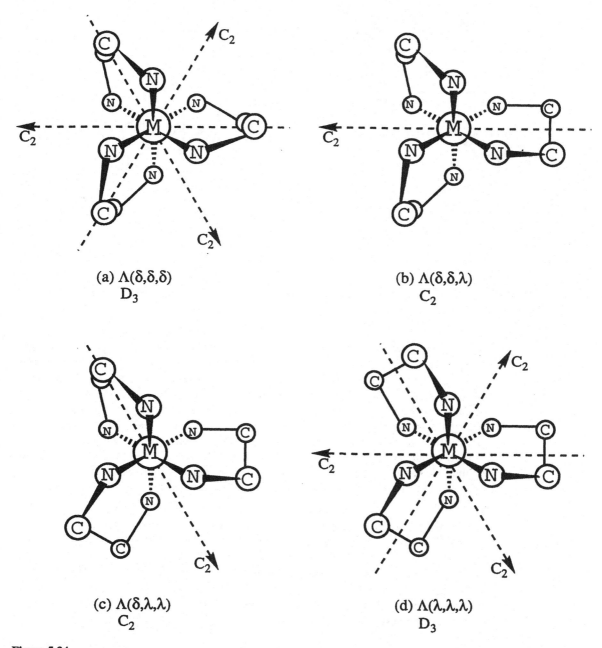

Figure 5.34
Projections of $\Lambda(\delta\delta\delta)$, $\Lambda(\delta\delta\lambda)$, $\Lambda(\delta\lambda\lambda)$ and $\Lambda(\lambda\lambda\lambda)$ for OC-6 [M(en)$_3$] complexes, along threefold or pseudo-threefold axes

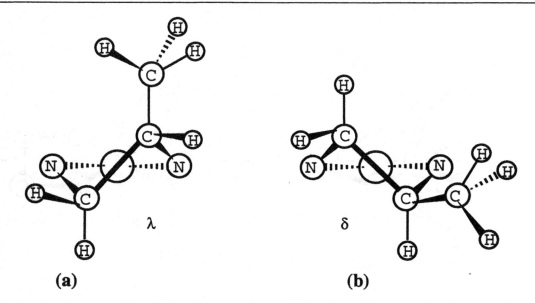

Figure 5.35
Equatorial (δ-configuration) and axial (λ-configuration) orientations of the methyl group in coordinated 1,2-diaminopropane

Conformational analysis is particularly pertinent in the case of ligands that have non-hydrogen substituents, e.g. 1,2-propylenediamine (pn). Of course, pn is a chiral ligand and it has to be considered also from this point of view in a complex. Here we want to discuss just the conformation of the chelate ring. There are two possible orientations of the methyl group, with respect to the chelate ring, if hydrogen atoms are replaced by CH_3 in Figure 5.35: equatorial (e) and axial (a).

It has been shown in a number of cases experimentally, and it is comprehensible on theoretical grounds, that the (e) preference is even stronger than it is in cyclohexane (Ref. 29, p. 686). One can therefore safely assume that the equatorial disposition persists in all pn complexes. This rule can be generalized to any kind of substituent.

Ligand conformations, six-membered chelate rings. Several complexes with six-membered chelate rings have been examined from the point of view of ring conformation in detail. In principle, the six-membered ring resembles the much studied cyclohexane ring of organic chemistry, except that the ligand bite angle is usually not far from 90°. Four possible conformations exist (Figure 5.36), e.g. for a tris complex with 1,3-diaminopropane: two rigid *chair* forms with mirror symmetry (designated as *p* and *a*, respectively) and two enantiomeric *twist-boat* forms with C_2 axes (λ and δ) [57].

The chair form of a six-membered chelate ring is not chiral. However, the interaction of such achiral rings with the metal center generates chirality. For *p* (parallel), the chair ring conformers fold in such a way that the central atom defines a rotation direction parallel to the direction defined by the metal ion configuration, and for *a* (anti-parallel) it is the inverse.

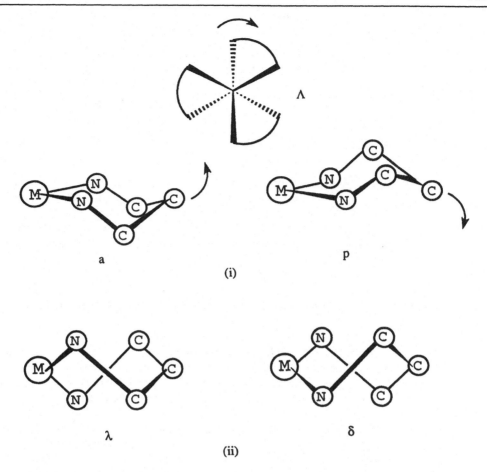

Figure 5.36
Two achiral chair forms (i) with mirror symmetry, designated as p (parallel) and a (antiparallel), respectively, and two enantiomeric twistboat forms (ii) with C_2 axes [57]

The boat form with mirror symmetry is not possible for tris complexes owing to strong inter-ligand interactions. It was shown [58] that in [M(1,3-pn)$_3$], 16 possible combinations of these conformers exist for each helical configuration Δ or Λ of the complex. Three are homoconformational and the others heteroconformational, with respect to the conformations of the three ligands. The energy differences between these conformers are relatively small, and similar complexes show different arrangements. Whereas the ligands in [Co(1,3-pn)$_3$]X$_3$(H$_2$O) (X=Cl, Br) are (chair)$_3$ homoconformational [59], the similar complex of Cr^{III} in [Co(1,3-pn)$_3$][Ni(CN)$_5$](H$_2$O)$_2$ has a heteroconformational arrangement (chair)$_2$(twist-boat) [60].

Conformations in metal complexes are not only pertinent for the coordination unit if they occur in the chelate cycle itself. In a series of ligands investigated by Thummel *et al.* [61] (5.10), the conformation of the ligand was examined as a function of the length of the aliphatic bridge.

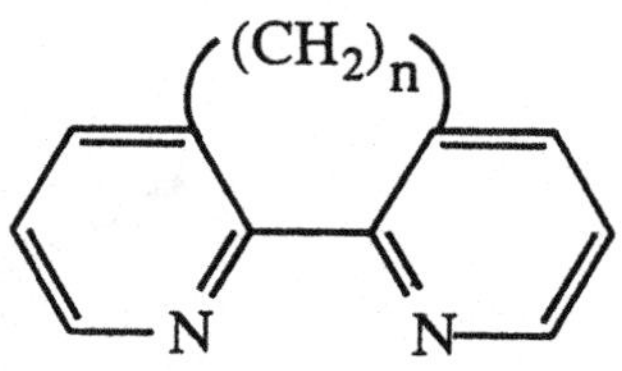

(5.10)

The conformation of the central ring depends strongly on the number of carbon atoms. For $n = 2$ and 3, the conformational changes are very labile, but for $n = 4$ (i.e. an eight-membered ring), the two pyridine units cannot be coplanar anymore. In a tris complex, a homoconformational arrangement is obtained, whereas four isomers are observed in [Ru(bpy)$_2$(3,3'-butylene-2,2'-bipyridine]$^{2+}$.

5.4   Coordination Units Comprising Non-cyclic Polydentate Ligands

Polydentate ligands form an unlimited set of molecules and a complete enumeration is therefore impossible. Some systematic surveyis given, however, since it facilitates comprehension and it can also help to inspire the preparative chemist to construct ligands that have special stereochemical features. A review by Bernauer [62] treats many aspects of diastereomerism and diastereoselectivity of polydentate ligands in a succinct way.

As a base, we want to consider ter-, tetra-, penta-, and hexadentate ligands, all themselves non-chiral, which have simple skeletons as those given in Figure 5.37.

For simplicity, we assume that the bridges between the ligating atoms are identical (e.g. two CH$_2$ groups). For the moment, we disregard the conformations of the chelate rings. The problem which we will address is the following: we consider complexes which comprise one of the above ligands, with the ligating atoms in one chelate ring coordinated in the *cis* position and we want to determine the number and the kind of isomers which exist. Again, we want to discus CN 4, 5, and 6, with a special emphasis on OC-6.

Coordination number 4

SP-4. Terdentate (A^A^A) and tetradentate (A^A^A^A) ligands can form SP-4 complexes, and numerous examples are known, especially for d^8 metals. The topographical stereochemical aspects of these complexes are relatively simple and no special features evolve from their consideration.

T-4. The small angle of 60° formed by the edges of a tetrahedron renders coordination of multidentate ligands unfavorable. Either the strain caused by this small angle is very large in the case of small chelate rings, or the chelate effect becomes small if larger rings are present.

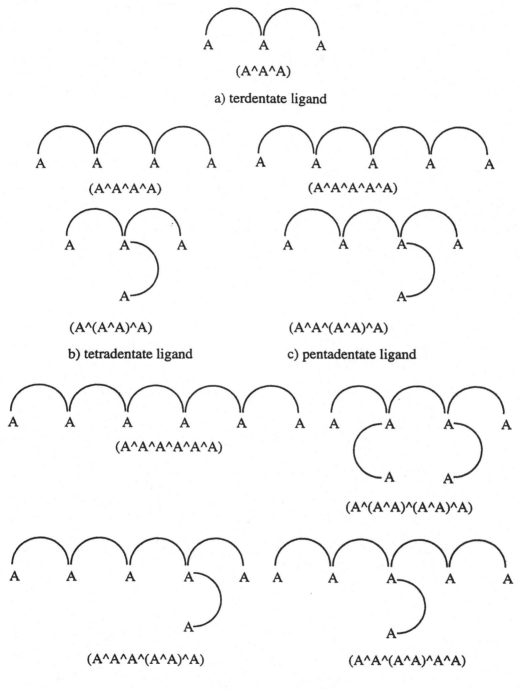

Figure 5.37
Schematic representations of some polydentate ligands

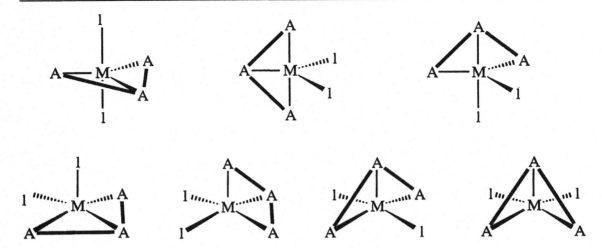

Figure 5.38
Possible arrangements for a terdentate ligand (A^A^A) in TB-5 and in SPY-5

Coordination number 5

A restricted number of complexes of the type $[M(A^\wedge A^\wedge A)(l)_2]$ is known and they are discussed in the frame of the repulsion model by Kepert [40]. In the idealized polyhedra TB-5 and SPY-5, three and four different arrangements are possible, respectively (Figure 5.38). They differ strongly in the angle between the two chelate rings.

The tetradentate ligand $[A^\wedge(A^\wedge A)^\wedge A]$ can stabilize the TB-5 coordination geometry through its tripod structure, and numerous cases have been reported.

A unique case of coordination geometry is provided by the V^{IV} center. It has a strong tendency to form as an inert unit oxovanadium(IV), or vanadyl cation VO^{2+}. The ligand obtained by fourfold deprotonation of H_4depa-X [63] provides an environment for the VO^{2+} ion (Figure 5.39), which results in highly stable complexes.

However V^{IV} does not always occur as oxovanadium(IV). It can form complexes with a normal OC-6 CG, e.g. with catechols [64].

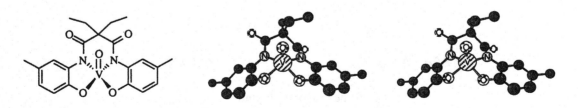

Figure 5.39
Example of a complex with CN 5 (4 + 1) with a VO^{2+} coordination center

Coordination number 6

The most investigated CG for polydentate ligands is again OC-6. There exist many possibilities of ligand arrangements, and interesting problems concerning stereoisomers occur. We want to discuss this case in a fairly general way. For that purpose it is useful to consider the complete set of *octahedral edge configurations*. An edge configuration is determined by a certain number N of octahedral edges, which are selected and which can be taken as a symbol for a chelate ring spanned across this edge. This problem has been discussed in detail in the literature [65]. There are for $0 \leqslant N \leqslant 12$ a total of 144 such configurations. Figure 5.40 contains the configurations for $0 \leqslant N \leqslant 6$. The configurations for $7 \leqslant N \leqslant 12$ can be derived immediately, by taking the so-called complementary configuration, i.e. the configuration, where the selection of edges (in our figure a bold edge) for $12 - N$ edges is inverted with respect to that with N edges. It is useful to draw all these edge configurations, and then proceed to examine their relationships with the coordination of actual ligands, such as those mentioned above. At this stage we disregard isomerism, due to the fact that amine donor groups can become stereogenic upon complexation to a metal. The consequences of this will be considered in Section 5.5. We also disregard the non-planarity of the chelate rings.

We shall number the edge configurations according to Ref. 65, and eliminate those which do not belong to the subject of our present discussion. Here we consider only non-cyclic polydentate ligands: 10, 18, 25, 29, 33, 34, 36, 37, 38, 43, 44, 45, 46, 47, 49, 51, 55, 56 and all configurations 58–87, and of course all ligands with more than six selected edges contain at least one cyclic structure. Of the remaining, 16 and 35 have a quaternary ligand atom, i.e. the ligand atom is connected with four other ligand atoms, a situation which is very unlikely to occur in a real case. Nos 1, 2, 5, 6, and 15 do not contain polydentate ligands and these configurations have already been discussed (1 has no chelating ligand, 2 one bidentate, 5 and 6 two bidentate, and 15 three bidentate ligands). Nos 9, 12, 13, 21, 24, 26, 27, 31, and 32 contain more than one chelating ligand. Nos 9, 12, 13, 21, 24, and 26 are heteroleptic, with respect to the chelate ligands, and we shall not discuss them further. All other configurations will be discussed below.

With terdentate ligands (A^A^A), two arrangements (3 and 4) are possible (Figure 5.41). Neither of them is inherently chiral, since the symmetry groups are C_s (3) and C_{2v} (4), respectively. Nos 27, 31, and 32 represent $[M(A^A^A)_2]$ complexes. Configuration 27 (C_1), gives a pair of enantiomers, whereas 31 has C_i and 32 D_{2d} symmetry.

Numerous complexes of configuration 32, e.g. Ru(terpy)$_2^{2+}$, are known [66]. Complexes of configurations 27 and 31 cannot be formed with planar ligands such as terpy. The non-rigid, aliphatic triamine dien (H$_2$NCH$_2$CH$_2$NHCH$_2$CH$_2$NH$_2$) forms all three isomers 27, 31, and 32 in the complex [Co(dien)$_2$]$^{3+}$ [67–69]. The TPY-3 geometry of the nitrogen centers in the free ligand, which becomes T-4 upon coordination, has some important consequences for the isomerism of the complex. Configuration 27 is chiral (C_2) by virtue of the chelate ring configuration, and it is therefore formed as a pair of enantiomers ($\Delta_2\Lambda/\Lambda_2\Delta = \Delta/\Lambda$). Configuration 31 is achiral (C_i). Configuration 32 represents an interesting case. As the edge configuration it is achiral (D_{2d}), and consequently complexes [M(A=A=A)$_2$] do not form enantiomeric pairs. However, the non-planarity of the

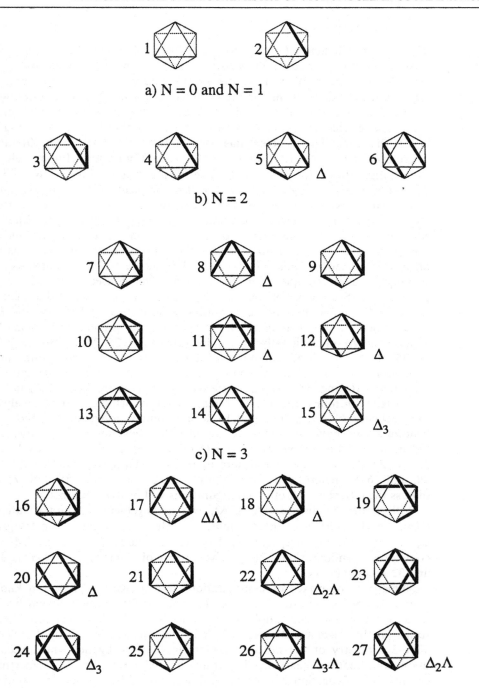

Figure 5.40
87 out of the 144 possible edge configurations of an octahedron [65]. The remaining 57 configurations are the inverses of configuration 1 to 57. The numbering scheme of this display is used throughout the book. For inherently chiral configurations the descriptors are given

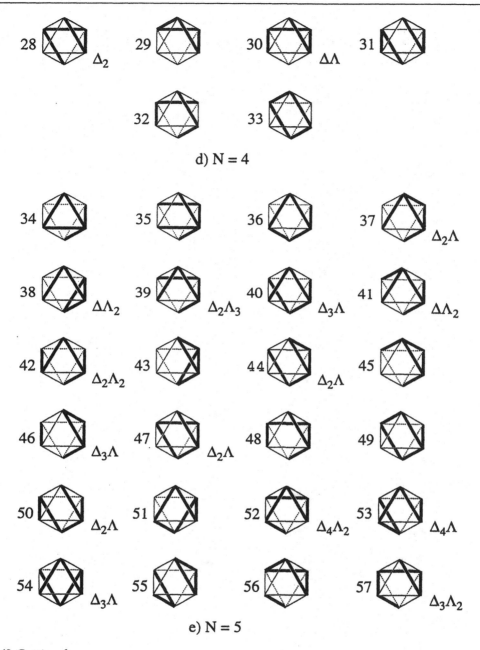

d) N = 4

e) N = 5

Figure 5.40 *Continued*

central ligand atom in the (A≈A≈A) type ligand dien breaks all mirror plane symmetries of the configuration, leaving a C_2-symmetric, chiral configuration (Figure 5.42).

Note that there is no stereogenic center created upon complexation. The two

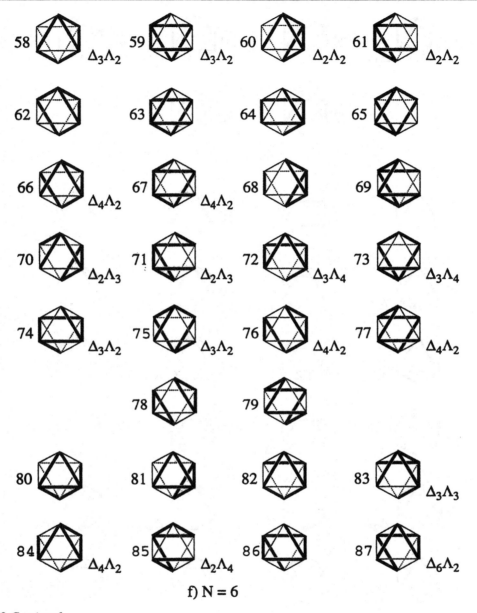

f) N = 6

Figure 5.40 *Continued*

N—H bond directions on the central secondary amine groups are two orthogonal lines and the Δ/Λ descriptors can therefore not be applied. The oriented line system for chiral reference can, however, be applied without problems. The enantiomers have been separated and they do not racemize rapidly [67].

Another type of a terdentate ligand is represented by 1,1,1-tris(aminoethyl)ethane (tame). this ligand can coordinate only in a facial configuration in OC-6 geometry.

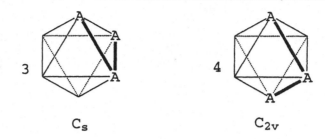

Figure 5.41
The two possible edge configurations for [M(A^A^A)] and their symmetries

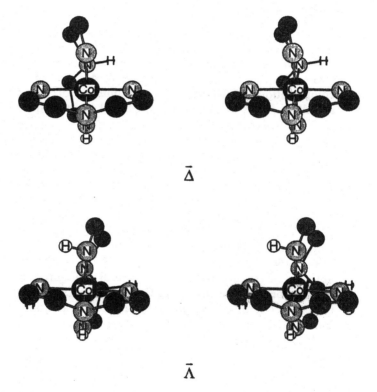

Figure 5.42.
Stereo pairs of the enantiomers of edge configuration 32 with the dien ligand

The 2 : 1 complex [Co(tame)₂]³⁺ was shown to have approximate D_3 symmetry in the solid state [70], indicating identical conformations in the two caps. A twist in one of the caps brings the symmetry to S_6, i.e. an achiral configuration. It is anticipated that the inter conversion is rapid in solution.

With tetradentate ligands, one non-chiral (C_s) configuration (7) is possible for [A^(A^A)^A] [Figure 5.43(a)]. Linear (A^A^A^A) has five configurations (two

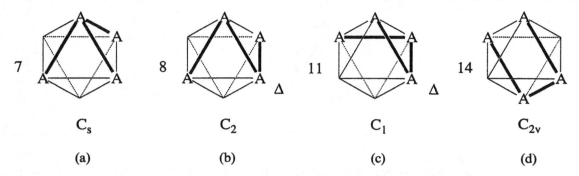

Figure 5.43
Edge configurations for [M(A^(A^A)^A)] and for [M(A^A^A^A)], with symmetries. The configurations 8 and 11, are often called a and b, respectively

pairs of enantiomers + one non-chiral configuration), namely 8, 11, and 14, if the non-planarity of the chelate rings is not considered [Figure 5.43(b,c,d)]. In fact, the two secondary amines of (A^A^A^A) become stereogenic centers upon complexation, which increases the number of isomers. We shall discuss these chirality centers in the section on chiral ligands. In the first report of complexes with the trien ligand [71], the configurations were designated as α-*cis* (8), β-*cis* (11), and *trans* (14). This nomenclature has later been used by many authors [72,73]. We prefer here either the numbering scheme of the edge configurations, or the classification by symmetries, which are C_2 (8), C_1 (11), and C_{2v} (14), respectively. Configurations 8 and 11 are chiral configurations, and the question arises as to which chirality descriptors should be used. This question is discussed at length in Ref. 65 and the reader is referred to this reference for further details. In the present case, the IUPAC rule that the terminal chelate rings have to be used for assigning chiralities suffices and gives Δ for both configurations indicated.

With the pentadentate ligand (A^A^(A^A)^A) five configurations (17, 19, and 20) are possible (Figure 5.44). The two chiral configurations (17 and 20) yield two pairs of enantiomers, whereas 19 has C_s symmetry. Configuration 17 is a special case with respect to chirality descriptors. It has been treated in [65] and given a Λ descriptor for the configuration depicted in Figure 5.44.

With the linear ligand (A^A^A^A^A), seven configurations (22, 23, 28, 30) are

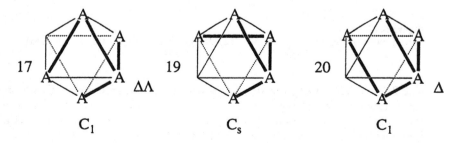

Figure 5.44
Edge configuration for branched pentadentate ligands (A^A˘(A^A)^A), with symmetries

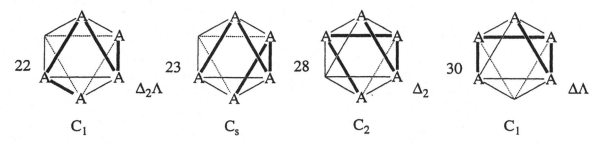

Figure 5.45
Edge configuration for the linear pentadentate ligand (A^A^A^A^A), with symmetries

possible (Figure 5.45). Three of them are chiral (symmetries: 22, C_1; 28, C_2; and 30, C_1) and can be assigned the chirality descriptors given in Figure 5.45. Configuration 23 has C_s symmetry.

The hexadentate ligand [A^(A^A)^(A^A)^A] has just two configurations, namely the pair of enantiomers of chiral, C_2-symmetric configuration 41 (Figure 5.46). The amine corresponding to this ligand type is the so-called penten (pentaethylenehexamine). The Co^{III} complex was prepared [74], their enantiomers were separated and their absolute configurations were determined [75,76].

(A^A^(A^A)^A^A) has two pairs of enantiomers from asymmetric configurations 39 and 40 (Figure 5.46). The ligand type (A^A^A^(A^A)^A) gives a total of five configurations, two pairs of enantiomers from asymmetric 42 and 50, and one C_s-symmetric form (48) (Figure 5.47).

Linear (A^A^A^A^A^A) (linpen) gives a total of eight configurations, namely four pairs of enantiomers from configurations 52 (C_2), 53 (C_1), 54 (C_2), and 57 (C_2) (Figure 5.48). All isomers have been isolated and their structures determined [77–79].

The ligands discussed above constitute a special selection, as the chelate rings are considered to have an equal sequence of atoms. There are a great number of polydentate ligands where different chelate rings are formed. As an example, we consider the case of EDTA. It forms one M(NCCN) chelate ring and four M(NCCO) rings. Since, from a symmetry point of view, it is equivalent to penten,

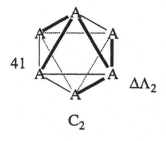

Figure 5.46
Edge configuration for dibranched hexadentate ligand (A^(A^A)^(A^A)^A), with symmetry C_2. Edta complexes have the same configuration

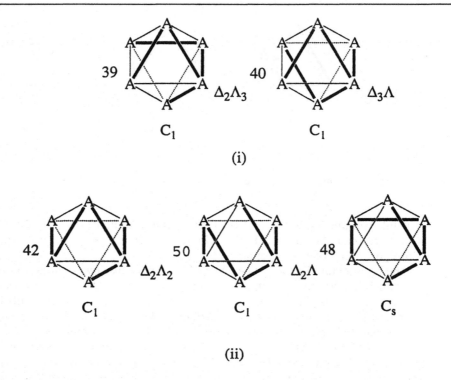

Figure 5.47
Edge configurations for monobranched hexadentate ligands (i) (A^A^(A^A)^A^A) and (ii) (A^A^A^(A^A)^A), with symmetries

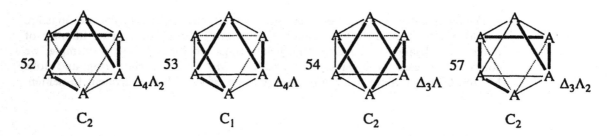

Figure 5.48
Edge configurations for linear hexadentate ligands (A^A^A^A^A^A), with symmetries

it is immediately apparent that it forms just one C_2 symmetrical complex of type (41). The use of the Bailar Tableau gives, as it must, the same result [23]. The structure, including the absolute configuration of one of the enantiomers of [Co(edta)]$^-$ [the (+)546 enantiomer], has been determined [80]. Its absolute configuration is $\Delta\Lambda\Delta$ and the NCCN chelate ring has λ conformation.

A linear hexadentate ligand with three different ligand atoms (O^N^S^S^N^O) was studied fairly long ago by Dwyer and co-workers [81–84] (Figure 5.49).

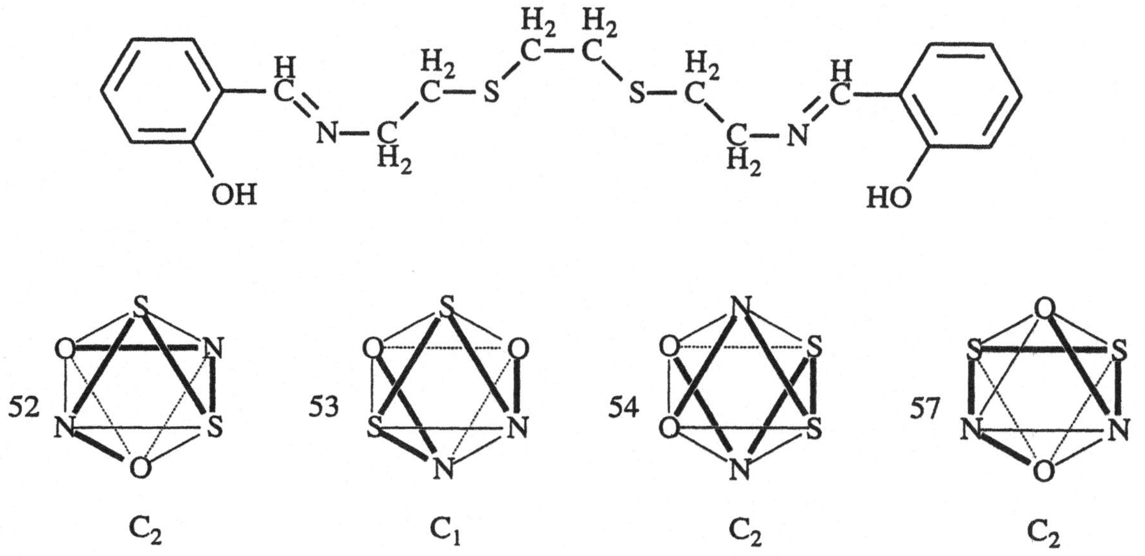

Figure 5.49
A linear ligand $O_2N_2S_2$ with four diastereomers of its M complex

As with the linear hexamine linpen, there are four pairs of enantiomers possible with this ligand. Only one pair of enantiomers, which were resolved, was observed, indicating a stereoselectivity introduced by the different ligating atoms [85]. The authors attributed the edge configuration 52 to the prepared isomer, basing their assumption on the fact that the double bond on the nitrogen is rigid and could adapt itself to the 90° angle more readily than to the 60° angle which is required for the other three diastereomers.

We shall make use of the edge configuration numbering scheme of an octahedron in several more cases, for complexes with other types of ligands, especially in Sections 5.6 and 5.7, since it is often helpful to trace a configuration back to a simpler case.

A linear hexadentate ligand of the type (A^B^^A^B^^A^B) is desferrioxamine B (DFO), a siderophore produced by *Streptomyces pilosus*, which is used therapeutically in cases of chronic iron poisoning [86]. It is a hydroxamic acid of structure (5.11). In this case, each of the four diastereomeric arrangements of (A^A^A^A^A^A) can occur in two forms. These diastereomers are depicted in Figure 5.50.

All these isomers are chiral, and appear therefore in true enantiomeric pairs since the ligand itself is achiral [87–89]. It was shown that the iron(III) ion uptake by *E. coli* is enantioselective. It prefers the Δ-absolute configuration of the metal complexes [90,91].

Polydentate ligands by no means always coordinate with all donor atoms. There are numerous examples where one or several 'arms' of a chelate are not bound to the metal center. One speaks of *pendant* chelating functions. EDTA, for example, forms complexes where only five, four, or two [92] of the six ligating atoms are coordinated.

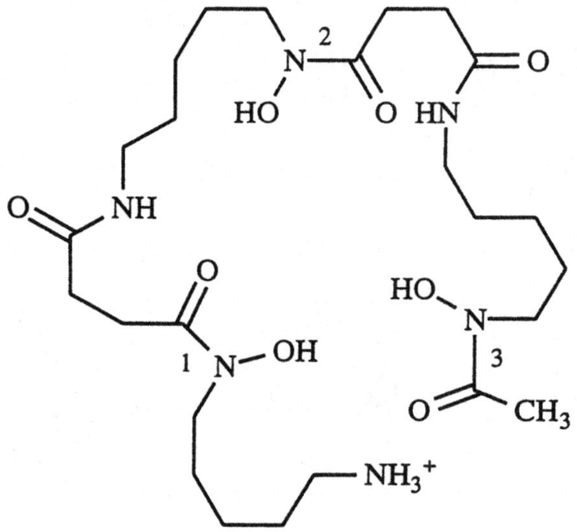

(5.11)

5.5   Coordination Units with Ligands Having Elements of Chirality

In this section, we want to discuss coordination units where the ligand segments themselves contain elements of chirality. The basis for this chirality has been termed 'vicinal elements' [93,94], yet this is not an especially appropriate designation and we therefore avoid it. One can distinguish basically three cases:

- One or more of the ligands coordinated to a metal are chiral entities by themselves.
- The ligand atom(s) itself can become a chiral center upon complexation.
- A symmetry, which renders a ligand containing stereogenic centers achiral, is broken upon complexation.

5.5.1   Inherently Chiral Ligands

The first coordination of a chiral ligand was described in 1907 by Chugaev. A large number of chiral molecules can coordinate to metal ions, and there are whole classes of chiral organic compounds with centers of chirality which form a wide variety of complexes. Another important class of chiral ligands are the C_2 symmetric atropisomers. Atropisomers (*a tropos*, not turning (Ref. 29, p. 1142)) are molecules having a chiral axis, owing to a hindered rotation around a single bond.

Chiral ligands may be present as racemates or as enantiomerically pure compounds

The largest class of chiral ligands, which are easily accessible as EPCs, are the naturally occurring α-amino acids[†] $NH_2CH(R')CO_2^-$. The absolute configuration

---

[†] R' is used to designate the substituent at the α-carbon in order to avoid confusion with the chirality descriptor *R*.

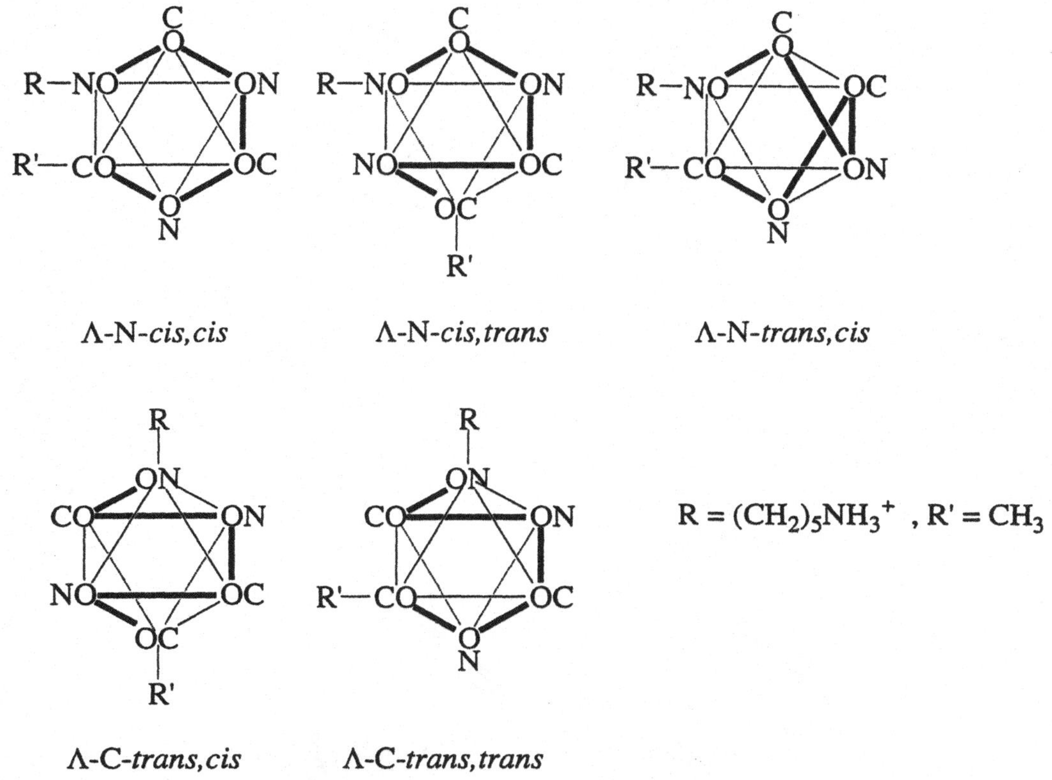

Λ-N-*cis,cis*   Λ-N-*cis,trans*   Λ-N-*trans,cis*

Λ-C-*trans,cis*   Λ-C-*trans,trans*

$R = (CH_2)_5NH_3^+$, $R' = CH_3$

Figure 5.50
The five diastereomers of desferrioxamine B (H$_4$DFO$^+$) complex with Cr^{III} [Cr(HDFO)]$^+$. Designations according to Ref. 87

can be either *R* or *S*, and accordingly we abbreviate these ligands as (A[*R*]^B) and (A[*S*]^B) for the enantiomerically pure compounds or (A[*R,S*]^B) for the racemic ligand. There is a wealth of information available on the coordination of these chiral ligands to metals, the most frequently investigated metal being cobalt in its +III oxidation state.

In any complex where the chelate ring configuration gives rise to a helical chirality, i.e. in complexes with two bidentate chelate rings leaving two coordination sites in *cis* positions, or in tris-bidentate complexes, different kinds of isomers occur with chiral ligands. Diastereomers arising from the occurrence of chirality elements in the ligands and in the first coordination sphere of the metal were isolated for the first time in the laboratory of Werner by Smirnoff. In one of his last publications, Werner reported the isolation of all eight isomers [Δ*R*/Λ*S*; Δ*S*/Λ*R*, each existing in two forms, see Figure 5.51(a)] of the *cis*-[Co(NO$_2$)$_2$(en)(pn)]$^+$ complex [95]. This work was later extended by Cooley *et al.* [96] to the similar, but simpler, complex *cis*-[Co(NO$_2$)$_2$(en)(bn)]$^+$, which yields four isomers [Figure 5.51(b)].

As a general example for an enantiomerically pure bidentate complex, we examine

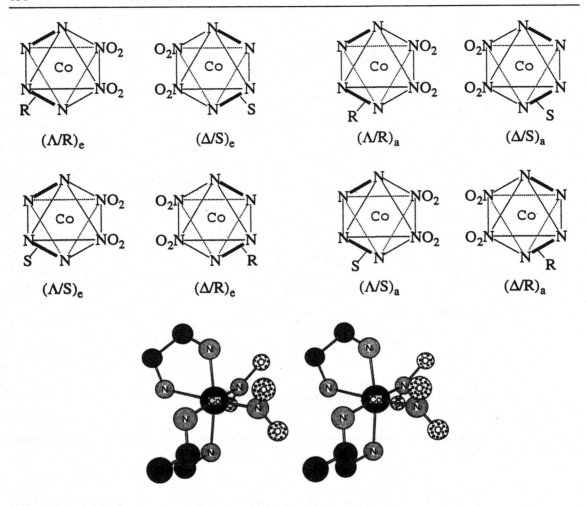

$(\Lambda/R)_e$ $(\Delta/S)_e$ $(\Lambda/R)_a$ $(\Delta/S)_a$

$(\Lambda/S)_e$ $(\Delta/R)_e$ $(\Lambda/S)_a$ $(\Delta/R)_a$

3D-representation of $(\Lambda/S)_e$

Figure 5.51
(a) Eight stereoisomers of $[Co(NO_2)_2en(pn)]^+$. Each enantiomeric pair exists in two diastereomeric forms that are designated as equatorial (subscript e) and axial (subscript a) with respect to the $\{Co(NO_2)$ plane. The $(\Lambda/S)_e$ isomer is also depicted as a stereo pair. (b) The four stereoisomers of $[Co(NO_2)_2en(bn)]^+$. The Λ/SS isomer is also depicted as a stereo pair

the isomers of $[M(A[R]^{\wedge}B)_3]$. There are two configurations, *mer* and *fac* for the rings, each with a Δ/Λ helical pair. Owing to the presence of the stereogenic centers, the two forms in a pair are no longer enantiomers, but diastereomers. A total of four diastereomeric complexes results (Figure 5.52).

Complex formation with a racemic ligand $(A[R,S]^{\wedge}B)_3$ brings for tris-bidentate complexes the total number of isomers to 16, i.e. eight pairs of enantiomers, not taking into account conformational isomers.

Since amino acids are usually used in their natural enantiomerically pure form,

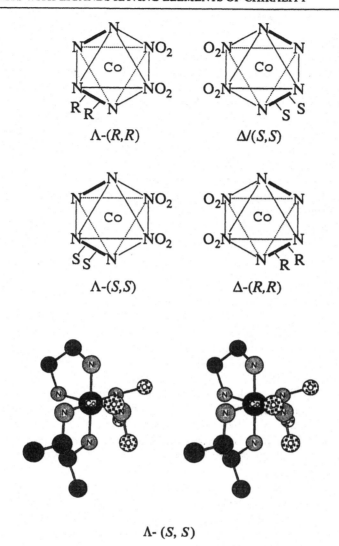

Λ-(S, S)

Figure 5.51 *Continued*

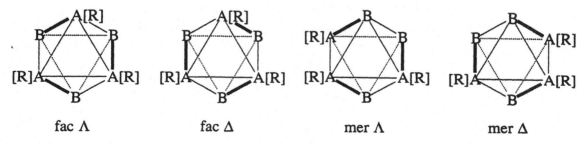

fac Λ fac Δ mer Λ mer Δ

Figure 5.52
Configurations of the four diastereomeric complexes [M(A[R]^B)₃]

the number of isomers obtained is much smaller than when racemic ligands are used. The absolute configuration of the helical arrangement of the ligands in the complex can immediately be assigned if a crystal structure is available, since the absolute configuration of the ligands themselves is known. For Co^{III} and Rh^{III} all isomers (Figure 5.53) in the series $Co(L\text{-ala})_3$ and $Rh(L\text{-ala})_3$ have been prepared [97,98]. Further important references concerning the stereoselectivity of complex formation with enantiomerically pure natural ligands are Refs 99–102.

A textbook example for inherently chiral ligands is the complex $[Co([R,S]\text{-}1,2\text{-propanediamine})_3]^{3+}$. Its stereochemistry has been studied in detail [103]. This complex represents 24 isomers (Table 5.8), which are due to a combination of configurational and conformational arrangements.

Note that the only conformations that allows the methyl group to adopt the much more stable (e) position are R_λ and S_δ. Several of the above-mentioned isomers have been isolated [104–106] and two structures were determined [107,108]. Equilibrated mixtures have been obtained at 100°C in the presence of charcoal (as catalyst), and they have been analyzed in order to establish free energy differences between the isomers [103].

While it is interesting and instructive to analyze the $[Co([R,S]\text{-}1,2\text{-propanediamine})_3]^{3+}$ system in detail, it demonstrates that racemic building blocks, as in this case the $[R,S]$-1,2-propanediamine, nearly always yield a large number of isomers which are generally not easily separated and characterized. The use of enantiomerically pure ligands reduces the number of isomers drastically. From $[Co([R]\text{-}1,2\text{-propanediamine})_3]^{3+}$ only four isomers are to be expected, instead of 24, namely *fac* and *mer* $\Delta(R_\lambda,R_\lambda,R_\lambda)$, and *fac* and *mer* $\Lambda(R_\lambda,R_\lambda,R_\lambda)$. In any kind of highly organized molecular assemblies, stereochemistry must be well defined. Nature demonstrates this very clearly by building up proteins exclusively from enantiomerically pure building blocks, e.g. the amino acids for proteins. If proteins were build from racemic amino acids, a *fuzzy stereochemistry* would result, which would not at all satisfy the requirements of the highly organized functionalities in biological systems.

Another case that has been investigated in detail are the tris 1,2-diaminocyclohexane (dach) complexes. Again, especially Co^{III} complexes have been investigated, but also Rh^{III}, Ir^{III} and Cr^{III} complexes are known [109].

It can easily be seen from the stereo pictures of the dach ligand that only the *trans*

Table 5.8. All stereoisomers of $[Co([R,S]\text{-}1,2\text{-propanediamine})_3]^{3+}$ due to different configurational and conformational arrangements

Conformation	Chirality	OC-6 coordination	No. of isomers
$(\text{lel})_3$	$\Lambda(S_\delta, S_\delta, S_\delta)$	*fac, mer*	2
	$\Delta(R_\lambda, R_\lambda, R_\lambda)$	*fac, mer*	2
$(\text{lel})_2(\text{ob})$	$\Lambda(S_\delta, S_\delta, S_\lambda)$	*fac, mer* (3)	4
	$\Delta(R_\lambda, R_\lambda, S_\delta)$	*fac, mer* (3)	4
$(\text{lel})(\text{ob})_2$	$\Lambda(S_\delta, R_\lambda, R_\lambda)$	*fac, mer* (3)	4
	$\Delta(R_\lambda, S_\delta, S_\delta)$	*fac, mer* (3)	4
$(\text{ob})_3$	$\Lambda(R_\lambda, R_\lambda, R_\lambda)$	*fac, mer*	2
	$\Delta(S_\delta, S_\delta, S_\delta)$	*fac, mer*	2
			Total: 24

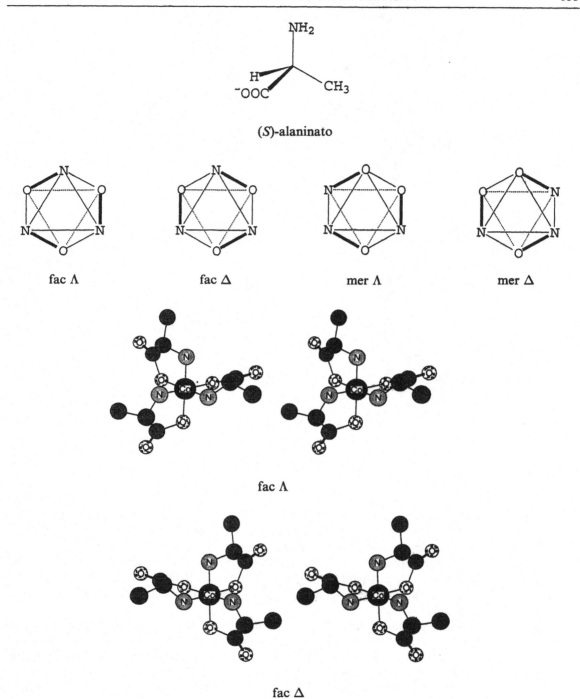

(*S*)-alaninato

fac Λ   fac Δ   mer Λ   mer Δ

fac Λ

fac Δ

Figure 5.53
Four diastereomers of [Co(ala)₃] [97]

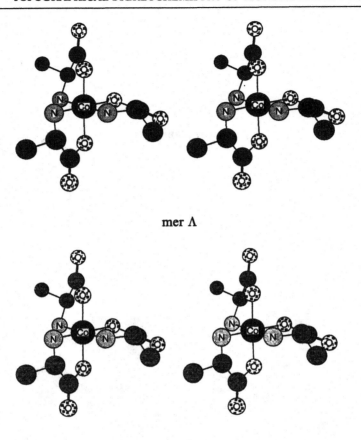

mer Λ

mer Δ

Figure 5.53 *Continued*

isomers with absolute configurations *R,R* or *S,S* can form a chelate with a metal (Figure 5.54). The conformations of these five-membered rings are fixed through the cyclohexane ring. Their configurations are λ for *R,R* and δ for *S,S*.

Enantiomerically pure ligands give therefore each two isomers, namely diastereomeric Δ/Λ-[Co(*R,R*-dach,λ)$_3$]$^{3+}$ and Δ/Λ-[Co(*S,S*-dach,δ)$_3$]$^{3+}$, respectively. All complexes have D_3 symmetry (approximate in the solid state owing to packing effects) [110].

With the enantiomerically pure ligands (either *R,R or S,S*), only one (ob)$_3$/(lel)$_3$ pair is obtained in unequal quantities, whereas with the racemic ligand, in addition to the four complexes mentioned above, the mixed complexes with $(S,S)_2(R,R)$ and $(S,S)(R,R)_2$ are also possible. In the latter case, a mixture of eight isomers, four of D_3 and four of C_2 symmetry (Ref. 111, p. 63) are possible. All eight isomers have been prepared [109] in solution and equilibrated at 100°C. From these equilibration studies relative free energy values were determined:

$$\Delta G° \text{ (lel)}_2\text{(ob)} \qquad \text{(lel)}_3 = -0.93 \text{kJmol}^{-1}$$

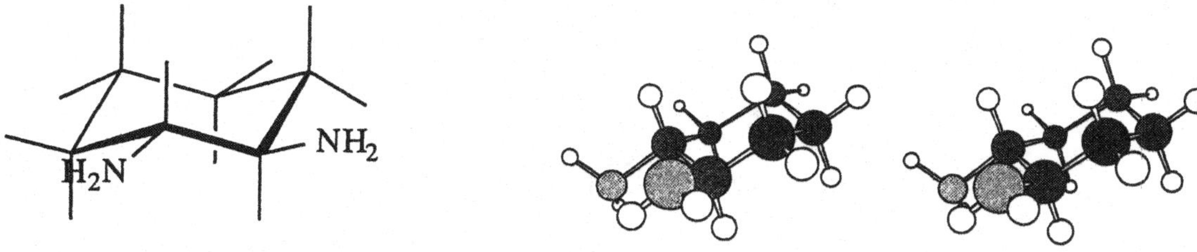

(*R*, *R*-dach)

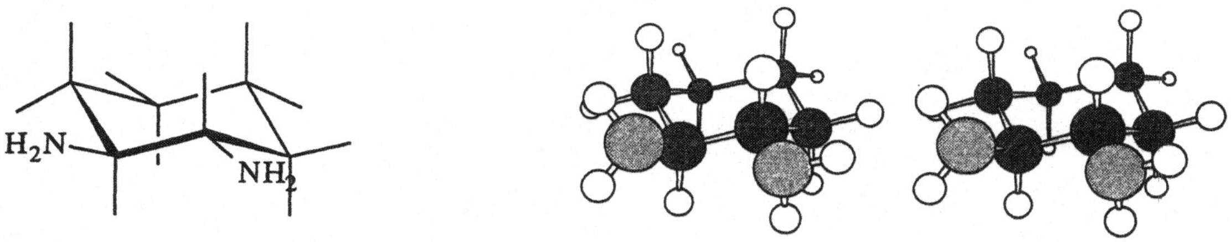

(*S*, *S*-dach)

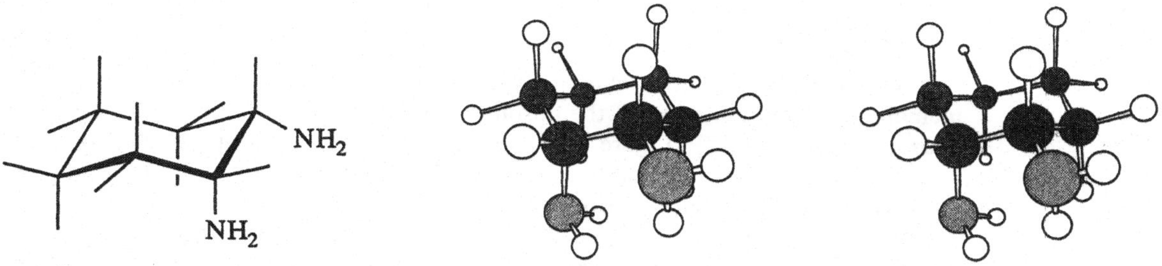

(*R*, *S*-dach)

$$\Delta G^\circ \text{ (lel)(ob)}_2 \qquad \text{(lel)}_3 = -3.72 \text{kJmol}^{-1}$$

$$\Delta G^\circ \text{ (ob)}_3 \qquad \text{(lel)}_3 = -8.20 \text{kJmol}^{-1}$$

These values indicate clearly the higher stability of the lel isomer.

Several chiral ligands, which show high diastereoselectivity in forming the two possible Δ and Λ forms of OC-6 complexes, have been described by Raymond and co-workers [112,113]. The ligand $H_2(PhMe)_2TAM$ (5.12), with homochiral S stereogenic centers, forms with Ga^{III} and with Fe^{III} the Λ configuration with a high selectivity. Iron(III), chromium(III), and cobalt(III) form diastereoselectively Δ or Λ complexes with chiral derivatives of 1-oxo-22(1H)-pyridinethionate.

The chirality in the ligand thus predetermines the metal chirality. Some further analogous cases are discussed in the literature [114].

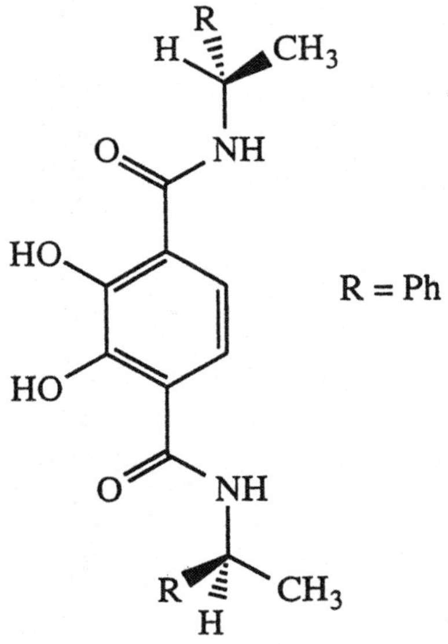

(5.12)

A detailed analysis of a chiral hexadentate ligand methylpenten (mepenten), including the determination of the absolute configuration of its Co^{III} complex, has been carried out [115] (Figure 5.55). Again, the methyl group is in equatorial position. It is interesting that an early conformational analysis of this complex predicted the conformations correctly [116].

The analogous complex with PDTA (propylenediamine teraacetate) had been studied by Dwyer and co-workers [117,118]. It was shown that this complex adopts the same configuration as the hexamine species [119].

An interesting naturally occurring terdentate chiral ligand is (*S*)-desferriferrithiocin (DFFT) (5.13) [120]. It acts in *Streptomyces antibioticus* as a siderophore. The OC-6 complex [M(*S*-DFFT)$_2$] can in principle appear in five chiral diastereomers that are not eliminated by the steric constraints of the ligand. All have C_2 symmetry, three of them are facial, and two meridional (both configuration 32) (Figure 5.56).

With Cr^{III} two complexes with meridional configuration are formed, both corresponding to the edge configuration 32. Although this configuration is achiral (symmetry D_{2d}), for symmetrical ligands (A^A^A) or (A^B^A), it becomes chiral at the coordination center, even for an achiral ligand (A^B^C).

The Δ/Λ chirality descriptors are not unambiguous in this case, since the parent edge configuration is achiral. The oriented line system yields an unambiguous designation. The *S*-configured carbon center in the naturally occurring ligand renders the two forms $\overrightarrow{\Delta}$-[M(A^B^C{*S*})$_2$] and $\overrightarrow{\Lambda}$-[M(A^B^C{*S*})$_2$], respectively, diastereomers. This is depicted in Figure 5.56.

Many chiral bi- and tridentate ligands have been prepared with the aim of generating catalytically active metal complexes for enantioselective reactions. A

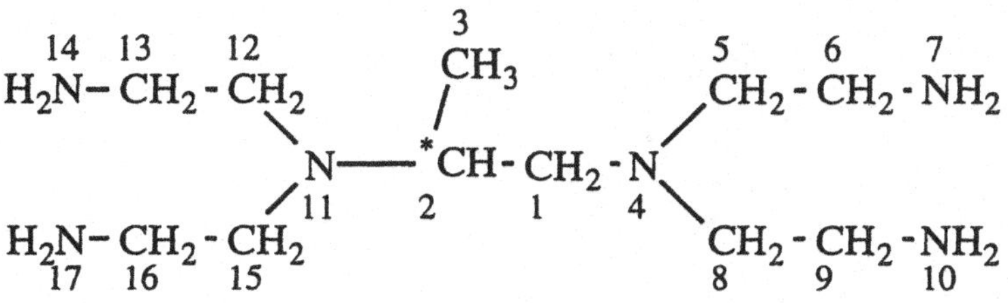

(i)

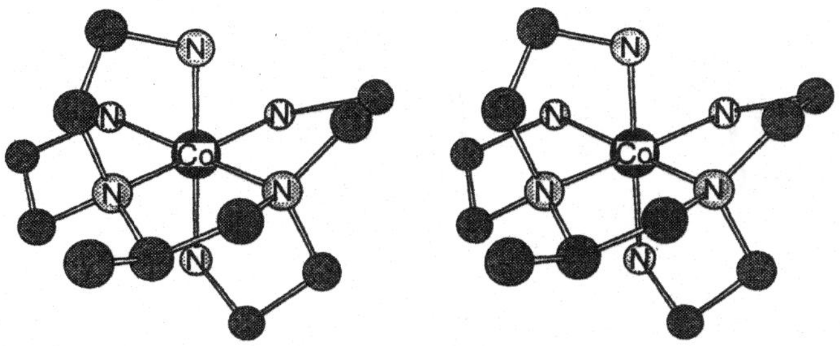

edge configuration #41 (Λ)
OC-6-34-C
(ii)

Figure 5.55
(i) The chiral ligand (R)-N,N,N',N'-tetrakis-(2'-aminoethyl)-1,2 diaminopropane = (R)-mepenten and (ii) the Co(III) complex $\Delta_2\Lambda$ (−)$_{589}$-[(R)–N,N,N',N'-tetrakis-(2'-aminoethyl)-1,2-diaminopropane]cobalt(III)

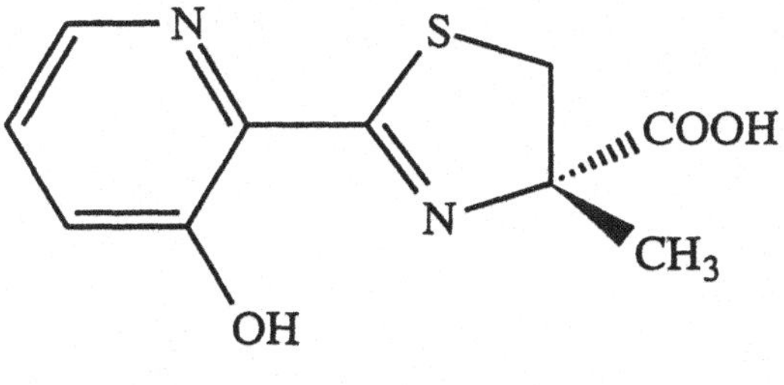

(5.13)

summary of such complexes and of the reactions involved will be given in Chapter 7, Section 7.6.

Recently a series of chiral tetradentate ligands were described [121,122] where chirality of the ligands is determined through preorganization of the helical chirality of the metal complex. With a general linear tetradentate ligand (A^A^A^A), three diastereomers (edge configurations 8, 11, and 14) can be formed, two of which (8 and 11) are chiral. The configurations can be preorganized by incorporating one or several stereogenic centers in the ligand in such a way that only one of the three diastereomers can be formed. If an enantiomerically pure ligand is used, the absolute configuration of the helical metal

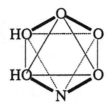

OH = phenolat oxygen
O = carboxylate oxygen

fac-N,N-trans, (N,O-Δ) (S,S)

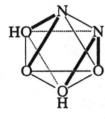

fac-N,N-cis, $\Delta_2\Lambda$ (S,S)   fac-N,N-cis, $\Lambda_2\Delta$ (S,S)

mer-(N,O-Λ) (S,S)   mer-(N,O-Δ) (S,S)

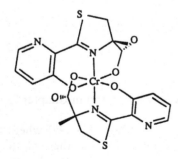

mer-(N,O-Λ) (S,S)

Figure 5.56.
Five diastereoisomers of a complex [M((S)DFFT)$_2$]. The two forms 31, Δ and Λ, were found in the Cr(III) complexes [120]

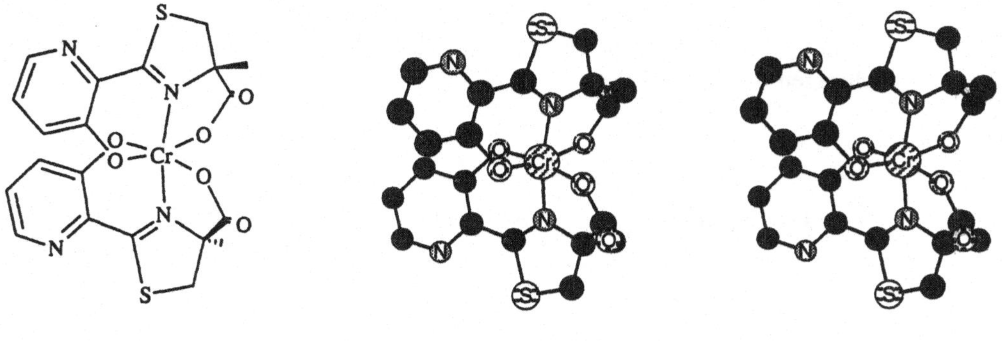

mer-(N,O-Δ) (S,S)

Figure 5.56 *Continued*

chirality will be predetermined. A symmetric ligand having stereogenic centers, which is synthesized from a racemic or from an achiral starting material will contain homochiral C_2 symmetric molecules of the racemate (A^A[R]^A[R]^A)/(A^A[S]^A[S]^A) and a heterochiral C_s symmetric *meso* form (A^A[R]^A[S]^A). By a judicious choice of the ligand, only the symmetry matching complex (ligand symmetry *and* edge configuration symmetry are both C_2) will be formed, yielding one of the racemates Δ-[M((A^A[R]^A[R]^A)]/Λ-(A^A[S]^A[S]^A) or Λ-[M((A^A[R]^A[R]^A)]/Δ-(A^A[S]^A[S]^A), depending on the detailed arrangements of the stereogenic centers in the ligand. If an enantiomerically pure ligand is used, only one stereoisomer will be formed.

A series of ligands fulfilling these conditions mentioned above are the '*chiragens*' [122] (5.14). The synthetic procedures allow for a wide choice of bridging units (B)

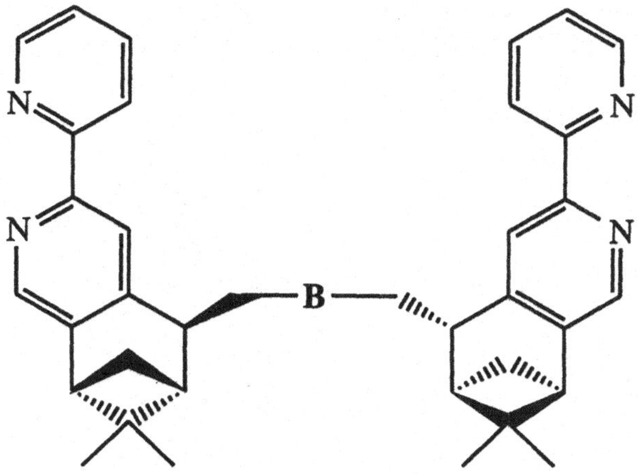

(5.14)

[†]The chiragen ligands generate helical chirality on the metal center on the basis of their central chirality on the ligands.

in the 'chiragen' ligands, providing the possibility of matching the coordination geometry of any given metal. Several bridges have already been realized with chiragen ligands (5.15).

B: H_2C—⬡—CH_2 H_2C—⬡—CH_2

H_2C CH_2

——$(CH_2)_n$—— $n = 0..6$

(5.15)

Many other bridging units can be introduced in this type of ligand.

The two halves of the C_2-symmetric chiragen ligands each contain three stereogenic centers, two already present in the pinene moiety and the third at the bridgehead. The latter is formed in a completely stereospecific way by the synthetic method used for the preparation of the chiragen ligand [121]. Figure 5.57 shows a computer-generated model of an Ru complex with a chiragen ligand that corresponds to the actual structure determined by X-ray crystallography.

Since the chiragen ligands are synthesized from chiral pool precursors, the ligands are obtained as enantiomerically pure compounds, or at least as enantiomerically enriched compounds, if the enantiomeric purity of the precursor is not 100%. The absolute configuration of the complex can be determined by an ordinary X-ray

Figure 5.57
Predetermined chirality at the metal center in the Ru-chiragen[6] complex. X-ray structure of the Δ-[Ru((+)-chiragen[6])(4,4′-dimethylbipyridine)]

diffraction method, since the absolute configuration of the stereogenic centers in the ligand is known.

Model considerations show that the 'mixed' ligands, i.e. that with R amd S at the two bridgeheads, cannot form complexes with an OC-6 central metal. This results in a selection in the complexation step where only the two homochiral ligands (R,R) and (S,S) show up in the metal complex. Through this selection process in the complexation step, a *chiral amplification* takes place if the precursor is not enantiomerically pure, which is often the case with pinene derivatives used for these syntheses. If the ligand synthesis starting from (A^A[R,S]) yields a statistical ratio of (A^A[R]^A[R]^A), (A^A[S]^A[S]^A), and (A^A[R]^A[S]^A), a simple calculation shows that a precursor with an enantiomeric purity ee will yield a complex with an enantiomeric purity $ee' = 2ee/(ee^2 + 1)$. The total yield y_t with respect to the ligand will be $y_t = 0.5(ee^2 + 1)$.

As will be discussed later, such complexes can be used as enantiomerically pure chiral building blocks for polynuclear species and also as ligands in enantioselective catalysis.

Diastereoselectivity in the formation of OC-6 complexes has also been observed with chiral linear N_4 ligands comprising two amine and two pyridine donors (5.16) [123]. Whereas (I) coordinates in a non-diastereoselective manner to Co^{III}, yielding [Λ-Co(^{III}(R-I)] with edge configurations 8 (C_2)(a-cis) and 11 (C_1) (b-cis), II yields one diastereomer only, which corresponds to edge configuration 8. This difference between I and II is obviously due to the difference in flexibility of the bridge between the two pyridine moieties.

(I)   (II)

(5.16)

Chiral pentadentate ligands which predetermine the helical chirality in OC-6 complexes have been described by Bernauer and co-workers [124,125] (Figure 5.58). These ligands coordinate in edge configuration 28, yielding a C_2-symmetric complex with predetermined absolute configuration, if the stereogenic carbon centers have the same absolute configuration.

Chiral C_2 symmetric ligands without stereogenic atoms (atropisomeric molecules) have become very important in enantioselective catalysis, and a number of cases will be discussed in that respect in Chapter 7, Section 7.7. As another example of a chiral ligand that has no stereogenic centers, we discuss here the complex [Ru(bpy)$_2$(1,1'-biiq)]$^{2+}$ (Figure 5.59) [126].

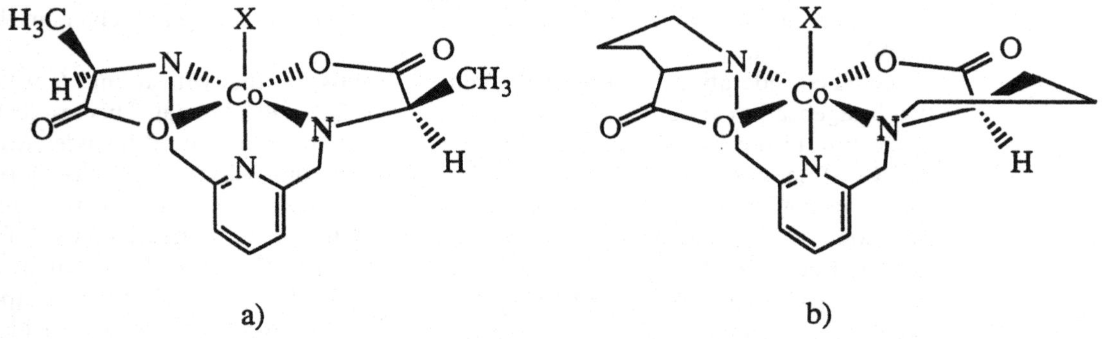

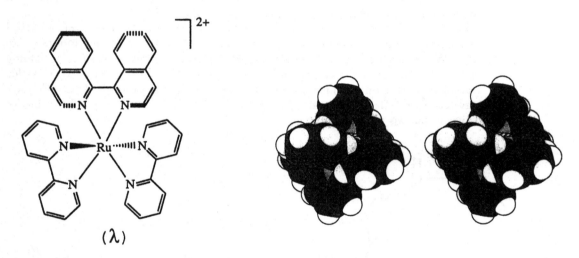

Figure 5.58
Predetermined absolute configuration of an OC-6 complexes with pentatentate ligands: a) Δ-[Co((*R,R*)-alamp)X]⁺ and b) Δ-[Co((*R,R*)-promp)X]⁺

Figure 5.59
[Ru(bpy)2(1,1'-biiq)]²⁺ [126] (1,1'-biiq = 1,1'-biisoquinoline)

The non-planarity of the 1,1'-biiq ligand introduces a chiral element into the ligand, to which the chiral descriptors δ and λ can be assigned. The complex shown in Figure 5.59 can therefore exist in four configurations, i.e. the two enantiomeric pairs Δ,δ/Λ,λ (A) and Δ,λ/Λ,δ (B), respectively. This situation corresponds to the conformational isomerism of the non-planar en ligand, the only difference being a much higher energy barrier for the δ → λ inversion (see Chapter 7, Section 7.1 for further discussion of the isomerization of this complex). The two pairs were found to exist in a 3 : 1 mixture (A : B) in thermodynamic equilibrium.

5.5.2 Ligands with Ligating Centers that become Stereogenic upon Complexation

TPY-3 ligand atoms in ligand molecules are often stereogenic and the free ligands exist therefore in two enantiomeric forms, but racemization is an extremely rapid process. This is especially the case for unsymmetrically substituted secondary or tertiary amine groups. Inversion of the TPY-3 nitrogen center usually takes place on a time scale of some GHz. Upon complexation, the TPY-3 center is transformed into a T-4 center, which is configurationally inert if the metal-to-ligand bond is inert. It is, of course, only configurationally stable as long as it is truly a T-4 center. If one of the ligands of the ligating N-center is hydrogen, deprotonation transforms the *coordinated* ligand into a TPY-3 center that is configurationally labile again. Deprotonation becomes therefore the controlling factor for configuration inversion.

Many complexes of Co^{III} and other metals forming inert complexes yield species with stereogenic ligand atoms. Given the rapid inversion in the free ligand, the ligands are always racemic before the complex is formed. A simple example is sarcosine and its complexes (Figure 5.60).

The complex [Co(sar)(en)$_2$]$^{2+}$ complex was studied very early [127] but detailed structural work carried out some 40 years later revealed that not all details of the original work could be confirmed [128]. The X-ray study in 1967 showed that the chelate ring is slightly puckered and the complete stereochemical characterization of one of the enantiomers of the most stable form is Δ-[Co(S-sar$_\lambda$)(en$_\lambda$)(en$_\delta$)]$^{2+}$. Subsequently, all four possible isomers have been prepared and the stability of Δ-[Co(R-sar)(en)$_2$]$^{2+}$ was determined to be 3.8kJmol^{-1} lower than the more stable diastereomer [129]. One pair of enantiomers is obtained in the tetramine complex [Co(R,S-sar)(NH$_3$)$_4$]$^{2+}$, which has been separated, and the absolute configuration determined [130,131].

Another early example of a complex with a ligand that forms a stereogenic center upon complexation is *N*-methyl-*N*-ethylglycine (Figure 5.61) [132].

The compound K[Pt^{II}(AMAC)(NO$_2$)$_2$] was resolved into the enantiomeric forms. Since the SP-4 CG of Pt^{II} excludes chirality of the metal itself, the successful separation is a proof of the non-labile stereogenicity of the non-deprotonable N-ligand.

The cases of secondary amine groups in the linear polyamine ligands discussed earlier are other examples of the fixation of chiral configurations of ligating atoms. As an example, we consider [M(A^A^A^A)]. If (A^A^A^A) is trien, the two

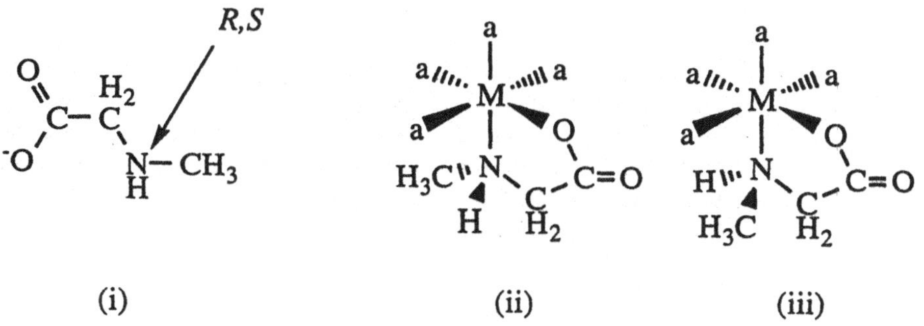

(i)   (ii)   (iii)

Figure 5.60
A ligands that becomes stereogenic upon complexation: (i) racemic sarcosine; (ii) two enantiomers of chiral [M(sarcosinato)a$_4$]

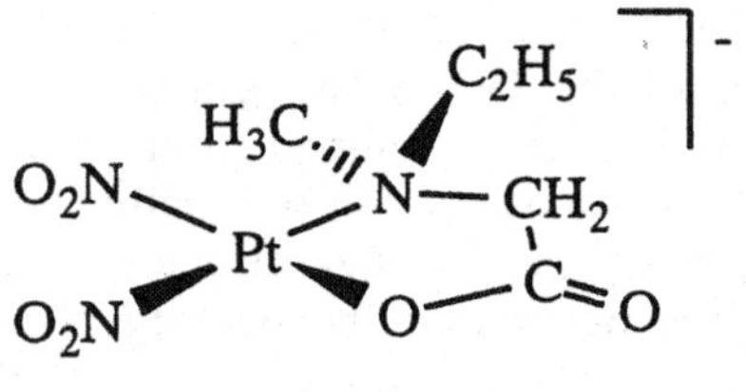

(i)   (ii)
(S)   (R)

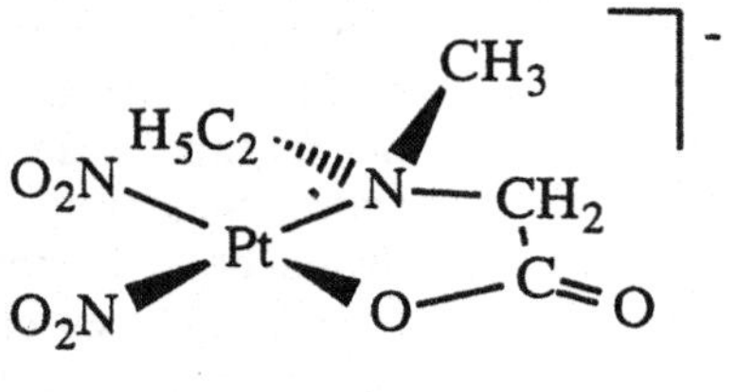

Figure 5.61
The enantiomeric pair of the SP-4 Pt(II) complex: N-methyl-N-ethylglycine
$[Pt(NO_2)_2AMAC]^+$ [132]

secondary amines become stereogenic upon complexation [133]. The detailed evaluation of the stereoisomers in this and similar cases is left as an exercise for the reader.

A particularly complete study has recently been published [134] dealing with complexes of the terdentate ligand glyphosate $[O_2CCH_2NHCH_2PO_3]^{3-}$, N-(phosphonomethyl)glycine (PMG) (5.17), which is the active ingredient in a commercial herbicide.

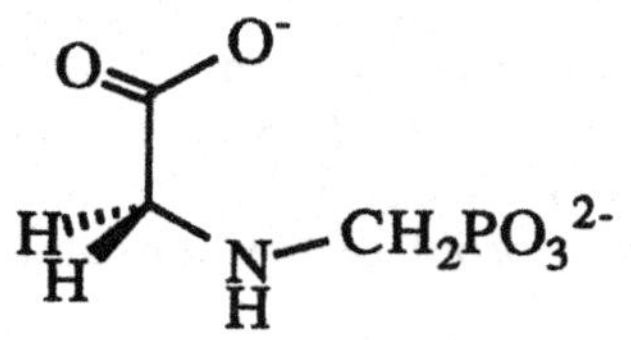

(5.17)

This ligand of the type $(A^\wedge B^\wedge C)$ has a TPY-3 nitrogen, which becomes a T-4 stereogenic center upon complexation. There are seven pairs of enantiomers and one achiral diastereomer in a $2:1$ OC-6 complex (Figure 5.62). All isomers have been found in solutions of the complex $[Co(PMG)_2]^{3-}$. The only achiral diastereomer, the *fac RS* all-*trans* complex (Figure 5.62) with symmetry C_i, crystallizes from an equilibrium mixture.

There exist, of course, ligands that have both stereogenic centers in the free form, and others, which become stereogenic upon complexation. An example is a $(P^\wedge N)$ ligand (5.18), which has a stereogenic carbon of known chirality (S) because the ligand is derived from valine, and the secondary amine which can become either S or R upon complexation [135].

5.5.3 Chirality due to Symmetry Relationships in Heteroleptic Complexes

There are cases where all ligands in a complex are achiral, despite the presence of stereogenic centers, i.e. the free ligand(s) is a *meso* form. If, in a heteroleptic

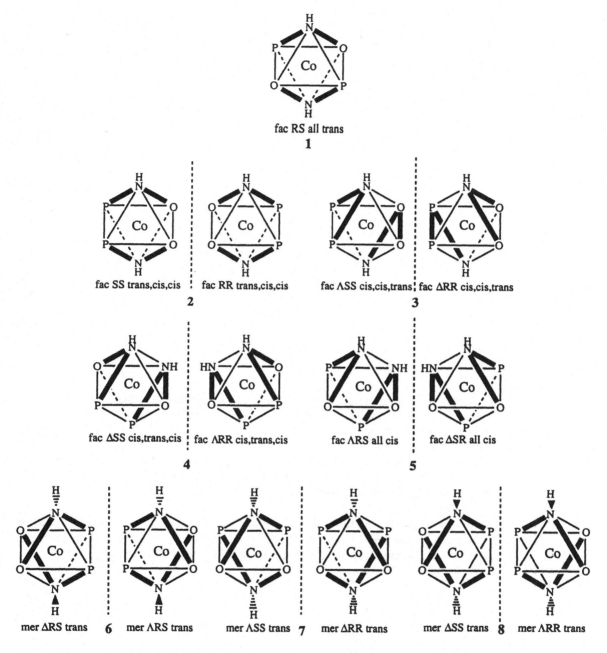

Figure 5.62
Schematic depiction of all possible $[Co(PMG)_2]^{3-}$ isomers. P represents the $CH_2PO_3^{2-}$ and O the $CH_2CO_2^-$ group

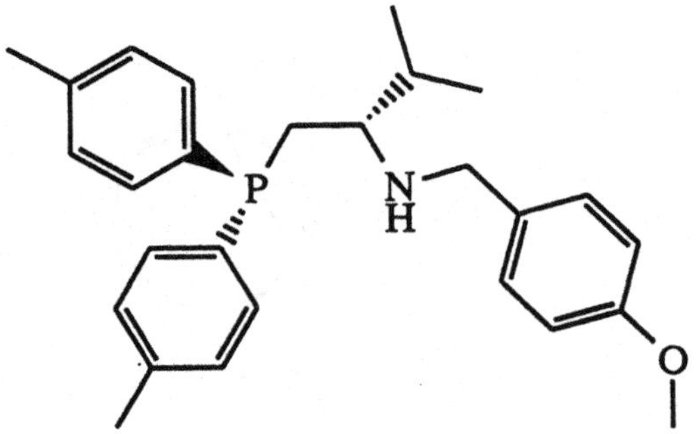

(5.18)

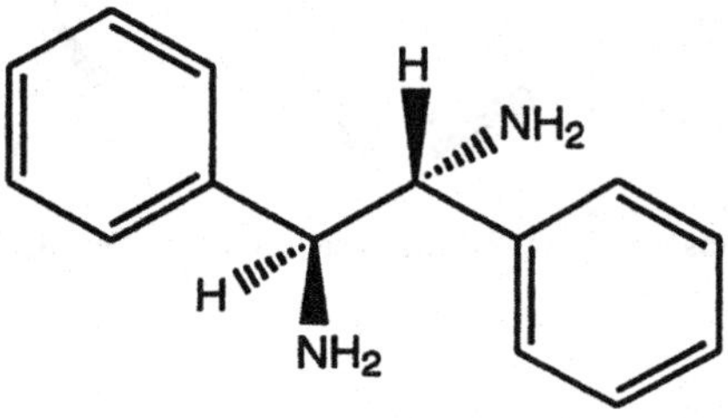

(5.19)

complex, the symmetry rendering such a ligand achiral is broken up by another ligand, a chiral complex may result. This type of chirality of a complex is not very frequent. An early and classical example is the SP-4 complex of Pt^{II} with one *meso*-stilbene diamine (5.19).

In the heteroleptic complex SP-4 [Pt(isobutylenediamine)(*meso*-stilenediamine)]$^{2+}$ (Figure 5.63), the mirror plane rendering the stilbene diamine a *meso* form, is broken by the second ligand. Thus the complex has symmetry C_1 and it is consequently chiral.

An analogous T-4 complex would have still C_s symmetry and it would therefore be achiral. The successful resolution of the Pt^{II} complex into a pair of enantiomers [136] was an elegant 'chemical' proof for the SP-4 coordination of Pt^{II}.

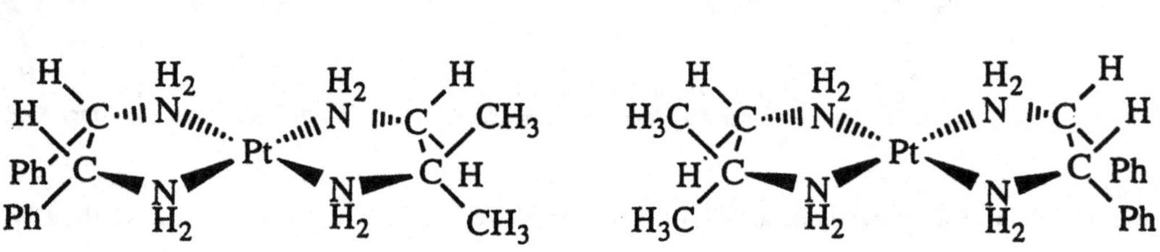

Figure 5.63
Enantiomers of SP-4 [Pt(isobutylenediamine)(*meso*-stilbenediamine)]$^{2+}$ [136]

5.6   Coordination Units Comprising Monocyclic Ligands

Cyclic ligands must have at least three donor atoms in a closed loop of atoms linked by covalent bonds. The smallest total number of atoms in the loop seems to be nine, and since classical organic chemistry dealt mostly with ring sizes ≤ 6, such molecules are often called macrocycles. There is a recent volume in the series *Stereochemistry of Organometallic and Inorganic Compounds* edited by Bernal about the stereochemical behavior of macrocyclic ligands, with three reviews on different aspects of this field [137–139] containing a lot of detailed information about such complexes.

In this section we shall discuss only monocyclic ligands, i.e. where there is only one possibility of making a complete loop following covalent bonds starting at the position of a donor atom. Polycyclic ligands will be discussed in Section 5.7. The donor atoms in macrocyclic ligands are in most cases O, S, N, or P. There are macrocycles, e.g. the crown ethers (see below), and the cyclam ligands, which have only one kind of donor atom, but there are also many molecules with two or three different donor atoms in the cycle. Macrocycles and their metal complexes have been known for a very long time from natural product chemistry, since two important metal complexes occurring in biological systems were relatively early identified as metal complexes of macrocyclic ligands. These are the heme group with iron as coordination center in molecules such as myoglobin and hemoglobin and the magnesium complex of chlorophyll. Both contain the porphyrin (5.20) macrocyclic ligand.

(5.20)

An analogous synthetic coordination unit, first obtained as the Cu^{II} complex [140] that can be synthesized in a relatively simple way, is the phthalocyanine ligand system (5.21). It has become the basis of a large industrial production of pigments [141].

A similar, yet distinctly different, macrocyclic ligand was found in the vitamin B_{12} molecule, the first natural product structure which was determined entirely by X-ray diffraction techniques [142]. The macrocycle that occurs in vitamin B_{12} is called the corrin structure (5.22).

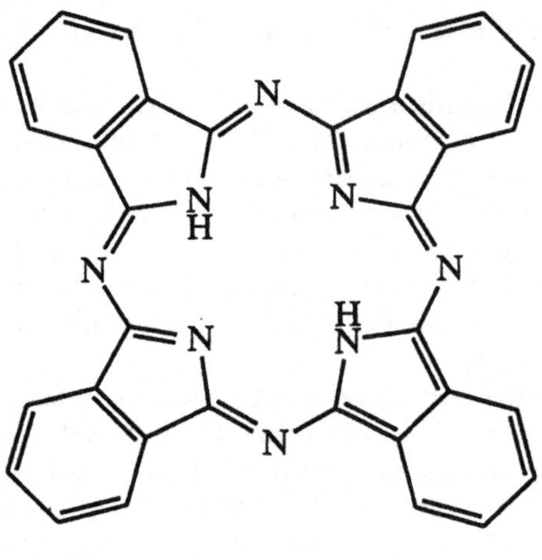

(5.21)

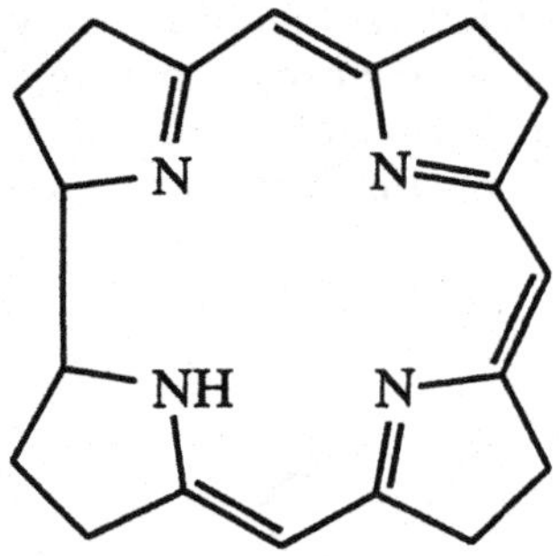

(5.22)

Both the porphyrin and the phthalocyanine structures consist of a 16-membered macrocycle where exclusively six-membered chelate rings occur. Two of the coordinating nitrogen donor ligands are deprotonated in the complexed form, rendering the ligands di-anionic. The corrin structure in contrast, provides a 15-membered macrocycle that forms three six-membered and one five-membered chelate ring. Only one nitrogen is deprotonated upon complex formation, rendering the ligand mono-anionic.

The common feature of these ring systems is their planar and rigid structure. This rigidity has far reaching consequences, since it contributes to a large extent to the kinetic and thermodynamic stability of these metal complexes. It is also the basis for important effects, such as the Perutz mechanism for the 'breathing motion' of the central iron atom in hemoglobin, which is responsible for the cooperative effect in oxygen binding.

Surprisingly recently, namely in 1960, the first synthesis of a 'floppy' macrocyclic ligand and its metal complex was reported [143]. This ligand, the cyclam (5.23), is an N_4 molecule synthesized in the ligand sphere of a transition metal ion.

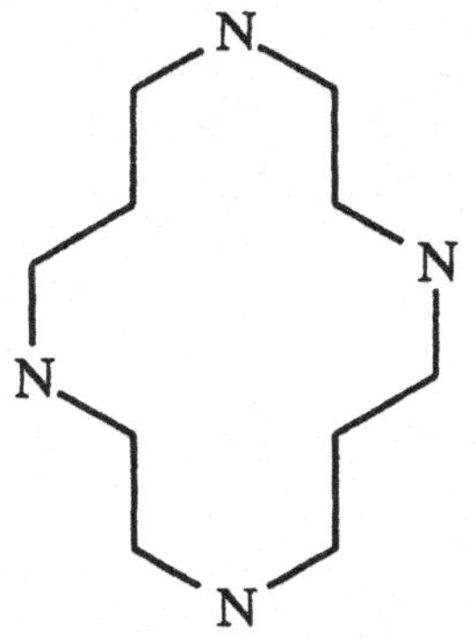

(5.23)

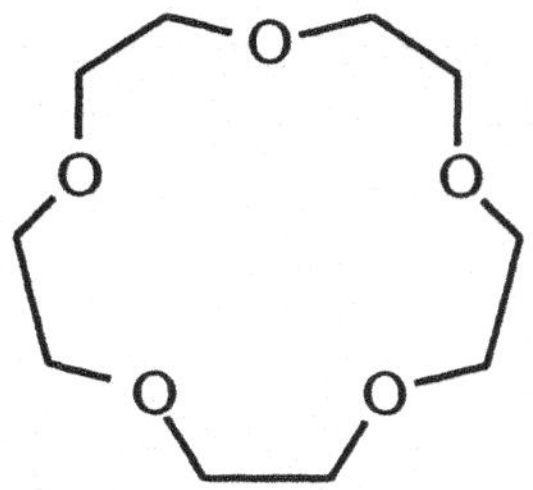

(i) 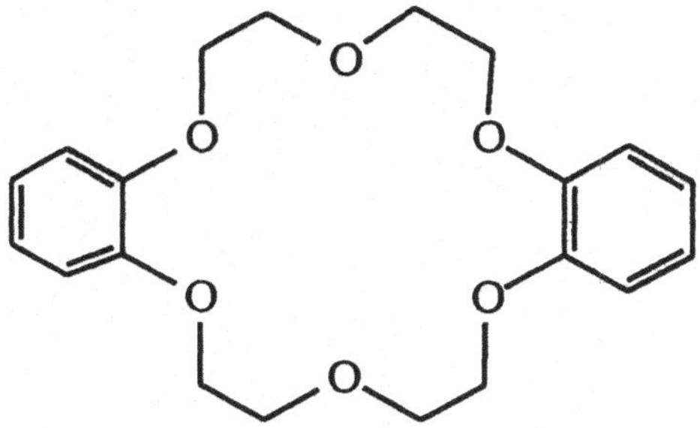   (ii)

(5.24)

Soon afterwards, Pedersen [144] reported the discovery of the alkali metal ion-binding properties of crown ethers, which triggered an enormous research activity in the subsequent decades. Two examples are given in (5.24)

The field has developed accordingly, and a large number of structures have been determined. The variety of structural details is too large to be covered in a general textbook about the stereochemistry of coordination compounds. For details, especially concerning conformational analysis of non-crown ether macrocyclic complexes, the reader is referred to Ref. 137; general treatments of macrocyclic complex formation can be found in several recent publications [145,146].

What was said about nomenclature in general (p. 61) is especially true for cyclic ligands and their metal complexes. IUPAC rules tend to yield extremely complicated names for such structures. It has therefore become customary to introduce *ad hoc* designations for certain groups of ligands such as the crown ethers. An often used shorthand nomenclature has been proposed by Melson [147]. This scheme is flexible and simple, so it can be adapted to actual cases rather easily. It consists of a number giving the ring size, a term denoting the unsaturation, and symbols of ligating atoms in alphabetical order. If there are substituents, they are put in front of the name, together with the positions where they are located. The numbering starts at the heteroatom according to CA rules, which give preference in the following order: $O > S > Se > N > P > As > Sb$, etc. Macrocycles having oxygen donors exclusively are called 'crown.'

The two examples given above obtain the following names: (5.24) 15-crown-5; (5.23) [14]ane-1,4,8,11-N_4. More examples for naming macrocycles will be given below.

5.6.1 Macrocyclic Complexes with N, S, and P Donor Atoms

We follow Ref. 137 and divide macrocycles into three groups: small-ring macrocycles, medium-ring macrocycles, and large-ring macrocycles. The distinction between these three groups is made according to the possibility that the metal is accommodated within the ring, if it is in a more or less planar conformation. In other words, if a metal does not fit into a macrocycle, it is called a 'small ring', if it just about fits into the macrocycle, the latter is called a 'medium ring', and finally it is called a 'large-ring' macrocycle if it has plenty of room in the cyclic structure.

Small-ring ligands. The most studied case is that of nine-membered rings, i.e. three donor atoms and three times two-carbon (CH_2 in most cases) bridging groups. These are the [9]ane cases (5.25).

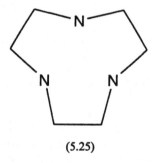

(5.25)

The majority of the complexes studied have OC-6 coordination geometry, mononuclear or with higher nuclearity (Chapter 6). Some other coordination geometries are also known. These are SPY-5 Cu^{II} complexes of composition $[Cu([9]ane-N_3)X_2]$, where X is a halogen, and a $[Pt([9]ane-N_3)_2]^{2+}$ complex with SP-4 coordination, leaving one N-donor per ligand uncoordinated. Both cases are manifestations of a strong tendency to avoid OC-6 coordination by these two metals. All other mononuclear complexes with this macrocycle seem to adopt the OC-6 geometry with the three nitrogen donors occupying a face of the octahedron. In many cases the conformations of the individual M—N—C—C—N chelate rings are (in the solid) identical, and they form a $(\delta\delta\delta)/(\lambda\lambda\lambda)$ racemate. In this case, a threefold axis passes through the center of the macrocycle and the metal.

Examination of the edge configurations of the octahedron, avoiding quaternary donor atoms, leaves as possibilities for complexes with this ligand type the arrangements 10, 18, 25, 37, 38, 43, 44, 45, 46, 55, 60, 66, 68, 70, 71, 76, 78, 83, 84, and 85. Some of them are chiral, others are not. Obviously, only a small number of

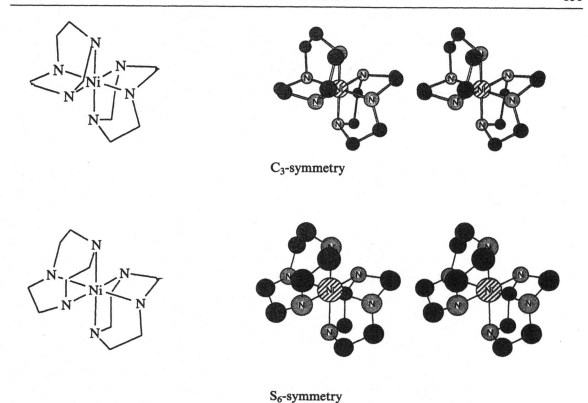

C_3-symmetry

S_6-symmetry

Figure 5.64
[M([9]ane-N$_3$)$_2$] with C_3 and with S_6 symmetry. Bis(1,4,7-triazacyclononane)nickel(II), in [Ni([9]ane-N$_3$)$_2$](NO$_3$)Cl · H$_2$O

these possibilities have been realized in the laboratory. Tridentate small ring macrocycles yield 1:1 complexes corresponding to configuration 10 and 2:1 complexes corresponding to 78. Both are inherently not chiral, except for the conformation of the ligand rings. In a 1:1 complex [M([9]ane-N$_3$)(l)$_3$], the highest possible symmetry (taking conformation into account) is C_3. In [M([9]ane-N$_3$)$_2$], the complex is either also C_3 if all chelate rings have the same conformational chirality, or S_6 if the chiralities of the opposite macrocycles are different. In actual cases [148] they seem to have approximate S_6 symmetry (Figure 5.64).

A derivative of [9]ane-N$_3$, TACTA, seems to have a strong tendency to impose TP-6 coordination geometry on certain metals, although [149] Cr^{III} and Ni^{II} clearly form OC-6 complexes. The edge configuration is that corresponding to 83, which is chiral. The descriptors Δ/Λ are applicable if only the pendant groups are considered. This ligand represents perhaps the simplest example of a class of coordination species where all ligand sites are connected in a way that the metal site is inside a one-sided open cavity. Such complexes, which can be termed 'half cage' structures, will be discussed in Section 5.7.

Sulfur analogues of the small-ring N$_3$-ligands have also been prepared and several

Figure 5.65
[Ni([10]ane-N_3)$_2$] from [152] in the compound bis(1,4,7-triazacyclodecane)nickel(II) diperchlorate, [Ni([10]ane-N_3)$_2$](ClO$_4$)$_2$

complexes were structurally characterized [150,151]. In the species [Co([9]-ane S$_3$)$_2$]$^{3+}$ the CG is distinctly OC-6.

Enlarging the macrocycle to a ten-membered ring gives, in the case of Ni^{II}, a complex [Ni([10]ane-N_3)$_2$] with C_s symmetry [152] (Figure 5.65).

Medium-ring macrocycles. These are mostly [12]ane and [13]ane ligands with three or four ligand atoms. These rings are too small to accommodate most metal ions completely within the ring, but often four ligand atoms in the cycle can coordinate to the metal. A [12]ane-P$_3$ ligand with threefold symmetry forms a C_3-symmetric complex with the Mo(CO)$_3$ moiety, occupying the face of an octahedron, in a very similar way to the [9]ane-N_3 ligands discussed above. [12]ane-N_4 ligands and their derivatives coordinate metals such as Co^{III} or Ni^{II} in a fashion that leaves two *cis* coordination positions for two monodentate ligands. This coordination corresponds to 29 (Figure 5.66) of the edge configuration scheme of Figure 5.40, an inherently non-chiral arrangement.

Large ring macrocycles. These are mainly [14]ane-D$_4$ ligands with various donor atoms D, the most numerous being those with D=N. The so-called cyclam ligand, which is [14]ane-1,4,8,11-N_4, was the first floppy macrocyclic ligand synthesized [143]. The cycle is sufficiently large to accommodate many metals in its 'flat' conformation, but it can also coordinate in OC-6 in a folded form, leaving two *cis* coordination sites occupied by a bidentate ligand. These two modes can be described by the edge configurations 33 and 29 (Figure 5.67) (Ref. 137, p. 51). Both configurations are (except for conformations) inherently non-chiral.

The cyclic structure can, in some cases, stabilize unusual oxidation states. Ag^I

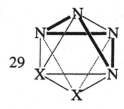

Figure 5.66
Edge configuration 29 for a mediumring macrocyclic ligand. The symmetry is C_s

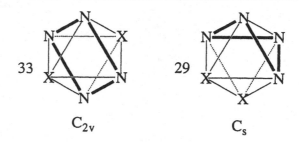

Figure 5.67
Edge configurations 33 and 29 for a larger medium ring macrocycle M([14]ane-N_4)X_2

disproportionates with cyclam to elemental silver and an Ag^{II} $4d^9$ complex. Interestingly, the latter shows two isomeric forms [153,154] (Figure 5.68), one in which the silver is displaced by 24pm from the plane defined by the four nitrogen donors, and a thermodynamically more stable form where the silver is coplanar with the four ligand centers. The former has C_s symmetry with the mirror plane bisecting the six-membered chelate rings, whereas the latter has C_i symmetry.

Variations of cyclam comprise C- and N-substituted cyclams. Examples are [14]ane-1,4,7,11-N_4 and [14]ane-1,4,7,10-N_4. A few complexes of these ligands have been characterized structurally [137]. S-analogs of cyclam have also been studied. The soft sulfur donor atom provides a favorable environment for soft metals such as Hg^{II}. An SPY-5 coordination has been found with Hg^{II} and ClO_4^- as anion, where

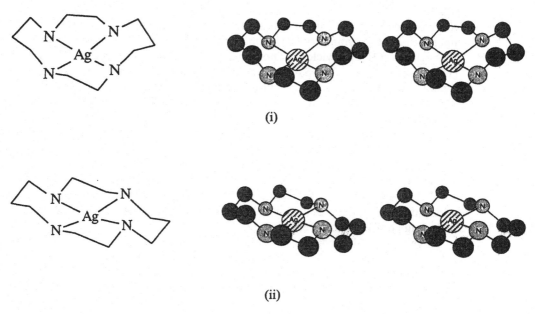

Figure 5.68
Two isomers of Ag^{II}-cyclam [153].

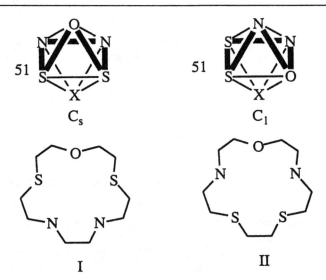

Figure 5.69
[15]Ane ligands in coordination with Ni(II). (I) [15]ane-7,10-N_2-1-O-4,13-S_2; (II) D(N^O)/
(S^S)-[15]ane-4,13-N_2-1-O-7,10-S_2

the Hg center is 48pm above the S_4 plane [155]. With this S_4 macrocycle, Ru^{II} forms a *cis* (edge configuration 29) complex with two additional Cl^- ligands.

[15]ane ligands have been prepared with four nitrogen donors, or mixed N_2OS_2, i.e. with five donor groups. Examples of the latter type are the two isomeric ligands [15]ane-7,10-N_2-1-O-4,13-S_2 and [15]ane-4,13-N_2-1-O-7,10-S_2. These two macrocycles form Ni^{II} complexes with edge configuration 51 (Figure 5.69). Both ligands have C_{2v} symmetry in the uncoordinated form. The ligand [15]ane-7,10-N_2-1-O-4,13-S_2 (N *cis*) yields a non-chiral C_s symmetric complex [156], whereas [15]ane-4,13-N_2-1-O-7,10-S_2 coordinates in such a way that the symmetry of the ligand is completely broken, resulting in a chiral configuration [157]. The absolute configuration of the complex can be assigned Λ (N^O)/(S^S), using the skew line reference system.

The 16-membered ring of 1,5,9,13-[16]ane-N_4 (5.26) resembles the ring system of the porphyrins and the phthalocyanines in its planar projection. It can accommodate metals in the ligand plane.

One 16-ring system with five N-donor ligating atoms in the ring, [16]ane-1,4,7,10,13-N_5, forms a Co^{III} complex with all N ligands bound to Co in a 51 configuration (Figure 5.40), which again has C_s symmetry, in this case even including conformation [158] (Figure 5.70).

Larger rings will be discussed with the crown ethers, where they are frequent and often investigated. With nitrogen donors, the cyclic [18]ane-N_6 ligand has in its Co^{III} complex an achiral 79-type configuration (Figure 5.40), where the conformations cause D_{3d} symmetry of the complex cation (Figure 5.71), which is consequently achiral.

Edge configuration 79 can be designated as *fac*, since all arrangements of three consecutive N-donors are facial. The only other possible edge configuration is 87

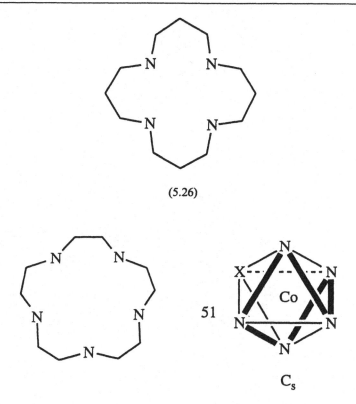

(5.26)

51

C_s

Figure 5.70
The edge configuration 51 that is found in a Co(III) complex with a pentadentate macrocyclic ligand.

(Figure 5.72). It can be called *mer*, because two triplets of consecutive N-donors (1, 2, 3 and 4, 5, 6) are meridional (Figure 5.72). This configuration is related to 32 of the bis-dien type, with the important difference that it is inherently chiral ($\Delta_6\Lambda_2 = \Delta$ and $\Lambda_6\Delta_2 = \Lambda$). The chirality of M(dien)$_2$ 32, caused by the T-4 configuration of the apical nitrogen donor atoms, is also present in 87 of [M([18]ane-N$_6$)]. This chirality can be conveniently described by the oriented line system $\overrightarrow{\Delta}/\overrightarrow{\Lambda}$. Consequently, in 87 two chiral diastereomers namely $\Delta(\overrightarrow{\Delta})/\Lambda(\overrightarrow{\Lambda})$ and $\Lambda(\overrightarrow{\Delta})/\Delta(\overrightarrow{\Lambda})$ are possible. All three diastereomers have been prepared [159,160].

5.6.2 Macrocyclic Complexes with Ligands Having Predominantly Oxygen Donors

In the wake of the discovery of the complex formation of alkali metals by crown ethers by Pedersen [144], enormous activity has arisen in research on this type of macrocycle. The basic structures of the crown ethers are cyclic structures with the composition (CH$_2$CH$_2$O)$_n$ ($n = 4$, 5, 6, 7). Following Pedersen's nomenclature, they are now generally called 12-crown-4 (12C4) (i), 15-crown-5 (15C5) (ii), 18-crown-6 (18C6) (iii), 21-crown-7 (21C7) (iv), etc. (5.27).

A large part of the work with crown ethers has focused on the 'selectivity' of a

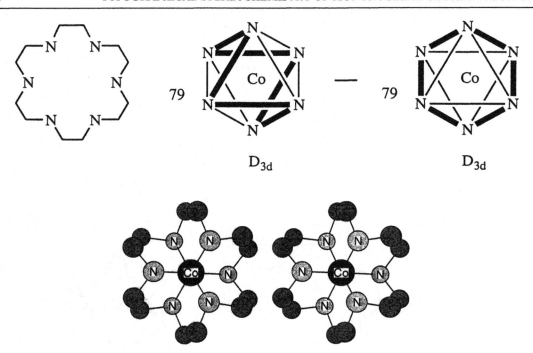

Figure 5.71
The complex $[Co([18]ane-N_6)]^{3+}$ in the D_{3d} diastereomeric form corresponding to the edge configuration 79 [159,160]

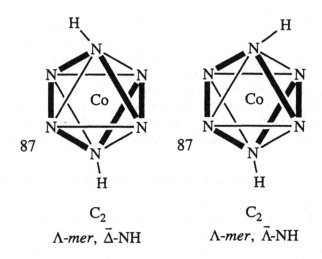

Figure 5.72
The complex $[Co([18]ane-N_6)]^{3+}$ in the two diastereomeric forms $\Lambda\text{-}mer/\overrightarrow{\Delta}\text{-}NH$; $\Lambda\text{-}mer/\overrightarrow{\Lambda}\text{-}NH$ of edge configuration 87. Each is one of a pair of enantiomers

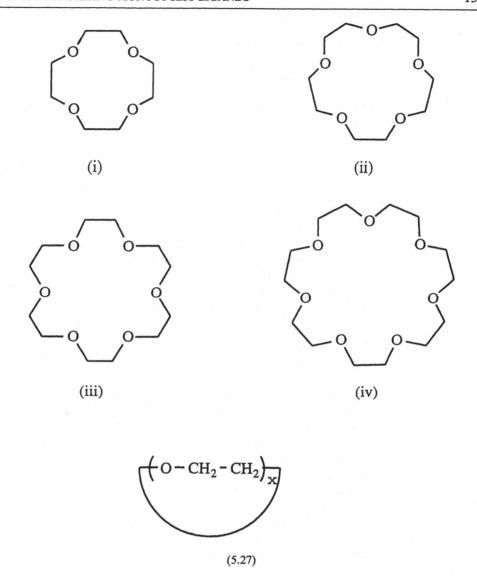

(i)

(ii)

(iii)

(iv)

(5.27)

given ligand for a series of metal ions. These oxygen donor ligands form complexes where the ligand–metal interaction is predominantly electrostatic, and the thermodynamic stability often largely determined by entropy factors [138]. 12C4 is to small to accommodate a metal ion completely, and 2:1 complexes can therefore occur. An example is the eight-coordinated complex with CG between SA-8 and CU-8 (the opposite faces of the cube are rotated through 30°).

Out-of-plane coordination seems also to be common for some of the larger rings, and large coordination numbers can result from 2:1 arrangements with these ligands. Electrostatic binding combined with high coordination number are favorable conditions for the formation of crown ether ligands with rare earth metal ions.

The synthesis and the properties of *chiral crown ethers* have been described by Stoddard [161]. Not many metal complexes with these interesting ligands have been studied yet.

5.7   Coordination Units Comprising Polycyclic Ligands. Cage Structures

Ligands in which the ligating atoms reside in a polycyclic structure yield metal complexes with fundamentally new properties. This is due to a basically different mode of dissociation (and of course a different mode of formation of the complex) of the ligand from the coordination center. Monodentate ligands can dissociate from a coordination center by a simple bond rupture (although such a process is generally accompanied by concomitant bond formation with an incoming new ligand), and chelate ligands can undergo detachment from the metal center through a sequence of such dissociation steps. Even in the case of monocyclic ligands, processes are conceivable where the dissociation of the ligand from the coordination center is a stepwise process, unless the ligand is very rigid. Polycyclic ligands, on the other hand, almost always require steps of simultaneous multiple bond ruptures for the dissociation, leading to very slow dissociation (and often also very slow formation) of the coordination units compared with complexes with similar 'open' ligating units. This highly increased kinetic and often also thermodynamic stability of the complexes has aroused the interest of many coordination chemists in the past 25 years, since such complexes have become synthetically feasible. Here we are primarily interested in the steric arrangements, and we shall therefore not discuss the intricate question how metals can get in and out of the inside of polycyclic ligands. Often central atoms can neither enter nor leave such structures at measurable rates under 'ordinary' conditions. The generally adopted name *cage ligands* is therefore highly appropriate. Cage ligands often have to be built through suitable reactions around the central metal.

The first cage ligands synthesized were the cryptands, which were designed for complexing alkali metal ions (5.28) [162].

Since alkali metal ions are spherical entities, the structures of the complexes are essentially determined by charge–dipole interactions. The thermodynamic and kinetic stabilities depend strongly on the cage size and numerous investigations

(5.28)

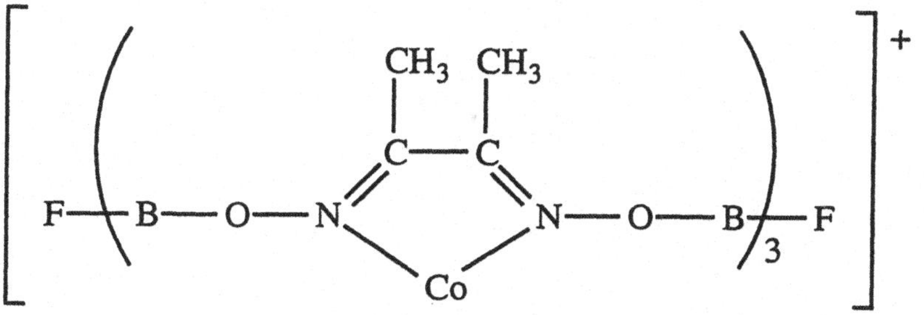

Figure 5.73
Cage complex with BF_3 capping [163] Clathro chelate derived from dimethylglyoxime, boron trifluoride, and Co(III). The bold lines in (ii) represent the chelate rings of the complex

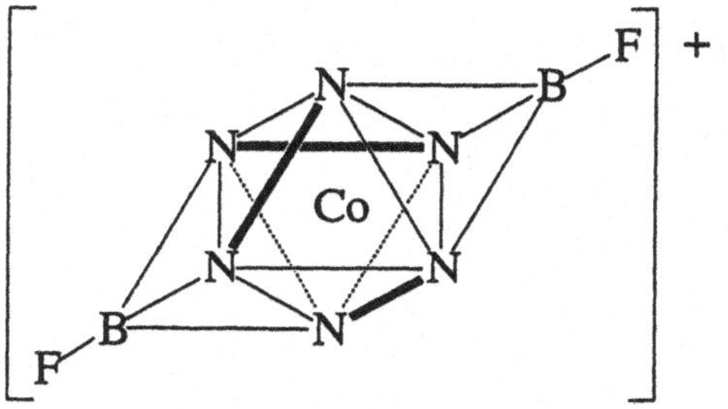

have been carried out to establish the factors governing the properties of such complexes.

Cage complexes of amine ligands with transition metals became accessible through the work of Boston and Rose [163], who showed that the tris(dimethylglyoxamato) complex of cobalt(III) reacts with boron trifluoride to the complex depicted in Figure 5.73.

Sargeson and his group in Australia [164] developed a versatile template synthesis for encapsulated metals and a large number of cage compounds of this type have been described since [165].

The cages [M(Sep)] and [M(Sar)] (Figure 5.74) are derived from [M(en)$_3$] complexes through reaction with CH_2O and NH_3 and CH_2O/CH_3NO_2, respectively, which results in a capping of the tris(bidentate) coordination unit. The [M(en)$_3$] unit is preserved during this reaction. There are two basic stereochemical questions involved in these cage compounds: (i) how large is the twist angle θ, determining the CG between OC-6 ($\phi = 60°$) and TP-6 ($\phi = 0°$), and (ii) what are the conformations of the NH—CH$_2$—CH$_2$—NH units in the complex? These two questions have been studied recently by strain energy minimization calculations [56,166]. The inversion of a single five-membered chelate ring of the en type has been calculated to be 21.6kJmol^{-1} in [Co(Sep)]$^{3+}$ [56]. Since the N-ligand atoms

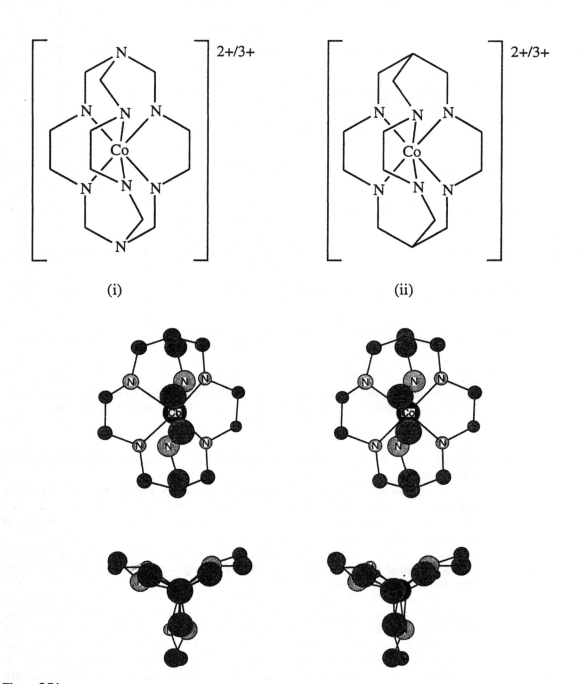

Figure 5.74
The D_3 symmetric cage complexes [M(Sep)] and [M(Sar)]: (i) $[Co(Sep)]^{2+/3+}$ and (ii) $[Co(Sar)]^{2+/3+}$ (schematic representations); (iii) stereo pair of Δ-$[Co(Sar)]^{2+/3+}$, (a) view along the C_3 axis and (b) view along the C_2 axis

become stereogenic upon coordination, the number of conformers in the cage is larger than in the [M(en)$_3$] unit. For example, the D_3 enantiomeric pairs $\Delta(\lambda,\lambda,\lambda)/\Lambda(\delta,\delta,\delta)$ and $\Delta(\delta,\delta,\delta)/\Delta(\lambda,\lambda,\lambda)$ of [M(en)$_3$], which can be designated as D$_3$(lel)$_3$ and D$_3$(ob)$_3$ respectively, (see p. 110) can exist in two diastereomeric forms in [M(Sep)] and [M(Sar)]: either the two caps are homochiral in which case the D_3 symmetry is preserved, or they are heterochiral with a concomitant reduction of the symmetry to C_3. A second conformational energy minimum, also with D_3 symmetry, has been identified for the D$_3$(lel)$_3$ case. It is called D$_3$(lel')$_3$ and differs from D$_3$(lel)$_3$ with respect to the torsion angles [166]. Experimental results [167] show that a variety of geometries occur. Table 5.9 gives the observed structures.

The synthetic method of bicapping an OC-6 complex can be extended to complexes with different ligating atoms on both sides of the cap, notably N and S (Figure 5.75) [168].

The structure of the complex [Co(azacapten)]$^{3+}$ has been investigated by X-ray crystallography in its ZnCl$_4^{2-}$/Cl$^-$ salt, as the racemate and as one of the enantiomers [(+)$_{510}$] (Figure 5.75). The former shows that the enantiomers are disordered in the crystal, thus forming a racemic solution. The enantiomerically pure complex has C_3 symmetry with all cysteamine rings being 'lel' conformed.

The absolute configuration of the enantiomer has also been determined. The stereochemistry of the complex can therefore be assigned unambiguously as (+)510-$\Lambda(\delta\delta\delta),\lambda_{Scap},\delta_{Ncap}$, where $(\delta\delta\delta)$ indicates the conformations of the five-membered cysteamine chelate rings, and λ_{Scap} and δ_{Ncap} the conformations of the caps on both sides of the coordination unit. An alternative way of indicating this stereochemistry is the assignment of the absolute configuration of the stereogenic ligand atoms in CIP nomenclature. Examination yields (+)$_{510}$-$\Lambda(\delta\delta\delta),\lambda_{Scap},\delta_{Ncap}$ = (+)$_{510}$-$\Lambda(S$-RRR,N-$SSS)$.

Another type of cage ligand is derived from catechol as the basic coordination unit. A series of complexes of Al^{III}, Ga^{III}, Fe^{III}, Ti^{IV}, and V^{IV} with the cage ligands BCT^{6-} and BCTPT^{6-} (5.29) have been systematically investigated [113], after it had been found that [Fe(BCT)]$^{3-}$ has TP-6 CG in its sodium salt [169–171].

The ligand BCT seems to be predisposed for structures with trigonal twist angles deviating considerably from the OC-6 value of 30° (see p. 106). The complex in Na$_3$[Fe(BCT)] has even genuine TP-6 CG (Figure 5.76).

A cage ligand of the catechol type with eight ligating atoms has been synthesized [172]. No structural information on a metal complex with this ligand is yet available. Models indicate the possibility of accommodating, e.g., Pu^{IV} in this cage.

According to Webster's Dictionary, a *cage* can be either an enclosure which is designed to confine permanently its contents, thus prohibiting escape of an object, but it can also be 'a frame with a net attached to it, forming the goal in ice hockey and field hockey.' Evidently, we want to see the puck going into this cage (at least if it is the goal of the 'others') and out again. We can speak of *closed* and *open cages*.

Open- or *half-cage* structures could also be called *octopus* ligands, if the etymology were extended from chelate (the pincers of a crab or a lobster) to another sea-dwelling animal. To our knowledge, no generally accepted designation for such complexes exists. Open-cage ligands seem to play an important role in some bioinorganic systems [113] and they represent interesting cases of designed metal complexes [173–176].

Table 5.9. Experimental structural parameters of transition metal hexamine cage complexes (from Ref. 166)

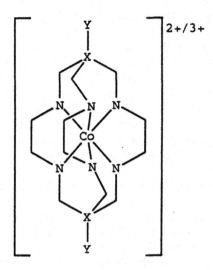

sar: $\qquad$ X=C $\quad$ Y=H
diamsar: $\qquad$ X=C $\quad$ Y=NH_2
di(amH)sar^{2+}: $\quad$ X=C $\quad$ Y=NH_3^+
di(NH_2OH)sar^{2+}: $\quad$ X=C $\quad$ Y=NH_2OH^+
sep: $\qquad$ X=N

Compound	Conformation	$M-N_{av}$ (Å)	Average twist angle (°)
Co^{III}sep(NO_3)_3	D_3lel_3	1.974	56.7
Co^{III}di(NH_2OH)sarCl_5 · 4H_2O	D_3ob_3	1.974	58.3
Fe^{III}sar(NO_3)_3	D_3lel_3	2.007	52.8
Cr^{III}diamsarCl_3 · H_2O	C_3lel_3	2.073	49.0
Ni^{II}di(amH)sar(NO_3)_4 · H_2O	D_3lel_3	2.110	47.1
Ni^{II}di(amH)sarCl_4 · H_2O	C_2lel_2ob_3	2.111	45.7
Ni^{II}sep(ClO_4)_2		2.111	48.0
Co^{II}sepS_2O_6 · H_2O	D_3lel_3	2.164	42.4
Cu^{II}di(amH)sar(NO_3)_4 · H_2O	D_3lel'_3	2.169	29.8
Co^{II}di(amH)sar(NO_3)_4 · H_2O	D_3lel'_3	2.170	29.0
Mg^{II}di(amH)sar(NO_3)_4 · H_2O	D_3lel'_3	2.188	27.8
Zn^{II}di(amH)sar(NO_3)_4 · H_2O	D_3lel'_3	2.190	28.6
Fe^{II}di(amH)sar(NO_3)_4 · H_2O	D_3lel'_3	2.202	28.6
Mn^{II}di(amH)sar(NO_3)_4 · H_2O	D_3lel'_3	2.238	27.6
Ag^{II}di(amH)sar(NO_3)_4 · H_2O	D_3lel'_3	2.286	28.8
Cd^{II}di(amH)sar(NO_3)_4 · H_2O	D_3lel'_3	2.30	27.4
Hg^{II}di(amH)sar(NO_3)_4 · H_2O	D_3lel'_3	2.35	25.8

Enterobactin (5.30), a siderophore produced by *Escherichia coli*, which has the largest stability constant yet known for Fe^{III} ($K = 10^{49}$ [177]), is a hexadentate open-cage ligand.

An interesting stereochemical feature of the enterobactin ligand is its inherent chiral nature. The capping part of the molecule is derived from the tri-l-serine

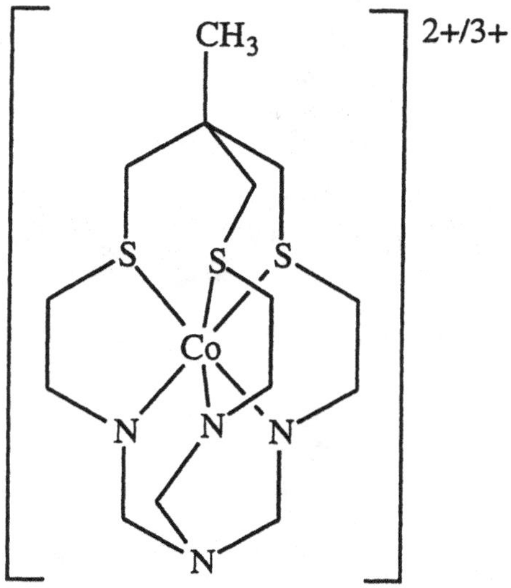

Figure 5.75
A cage complex with C_3 symmetry, $[Co(azacapten)]^{2+/3+}$ (Ref. 168, p. 2704)

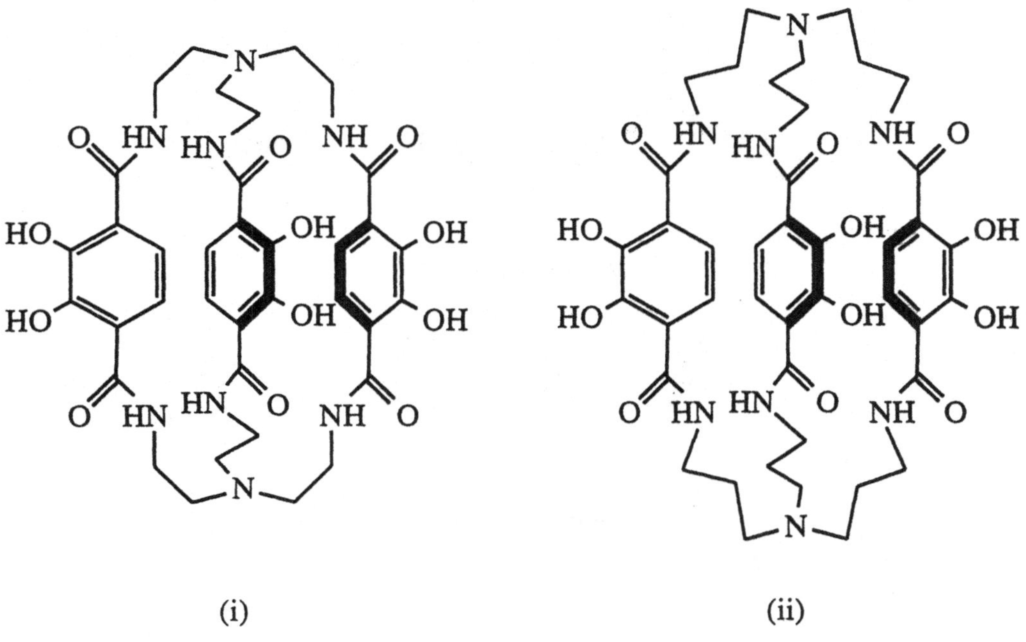

(i)

(ii)

(5.29)

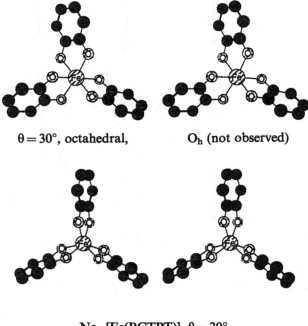

$\theta = 30°$, octahedral, O_h (not observed)

$Na_3 [Fe(BCTPT)]$, $\theta = 20°$

$K_3 [Fe(BCT)]$, $\theta = 5°$

Na$_3$ [Fe(BCT)], $\theta = 0°$, trigonal prismatic; D_{3h}

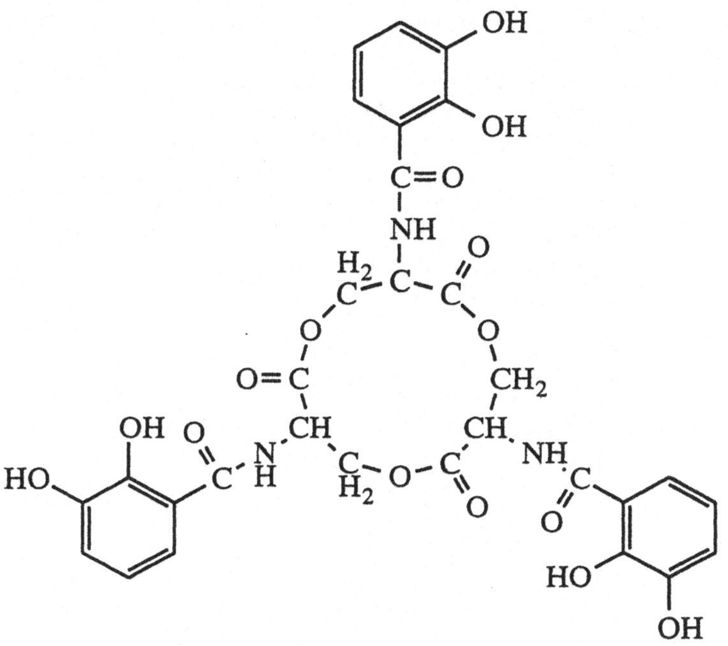

(5.30)

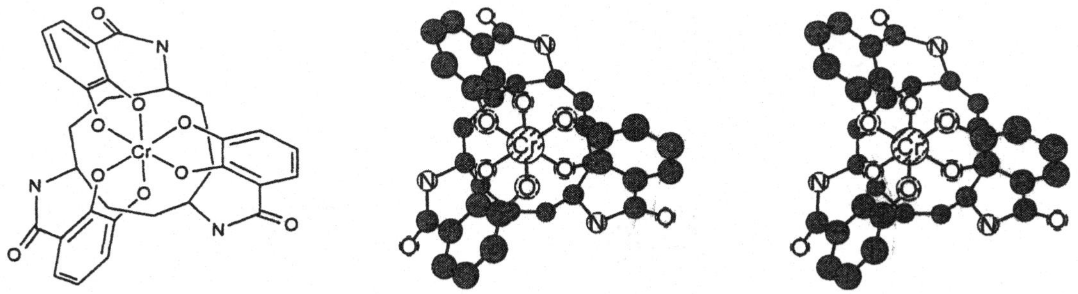

Figure 5.77
Stereopictures of a Λ-Cr-(SSS)-enterobactin complex, viewed along the C_3 axis

lactone and has S-chirality for all three stereogenic centers. The helical chirality of this distinct OC-6 complex (as opposed to TP-6; see also p. 161) seems to be completely predetermined. Δ and Λ are, of course diastereomers, and the complex seems to be formed in a highly diastereoselective manner [113]. The Δ,Λ diastereoselectivity of enterobactin was also shown for the complex $[Cr^{III}(ent)]^{3-}$,

Figure 5.76 (*Opposite*)
Six complexes with the BCT ligand, viewed along the threefold axis. $\Theta = 0°$ corresponds to TP-6, $\Theta = 30°$ to OC-6

which has Δ-configuration for the natural (*SSS*)-form of enterobactin (Figure 5.77) [178].

Ferrioxamine analogs, derived from natural enantiomerically pure amino acids, have been examined for their chiral preference in the binding to Fe^{III} [179].

Cage and octopus ligands with bidentate coordinating units of the bipyridine type and the corresponding complexes have been reported [173,180–182]. Such ligands are powerful sequestering agents for metal ions. An example is given in Figure 5.78.

A hexadentate octopus ligand [183] was especially designed to complex the UO_2^{2+} ion, which has HB-8 (hexagonal-bipyramidal) CG (Figure 5.79). Here the six

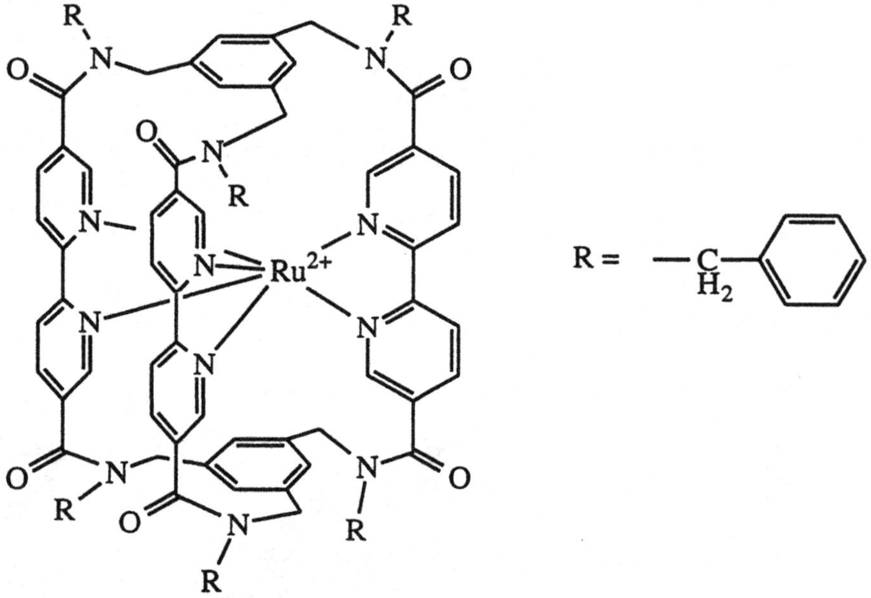

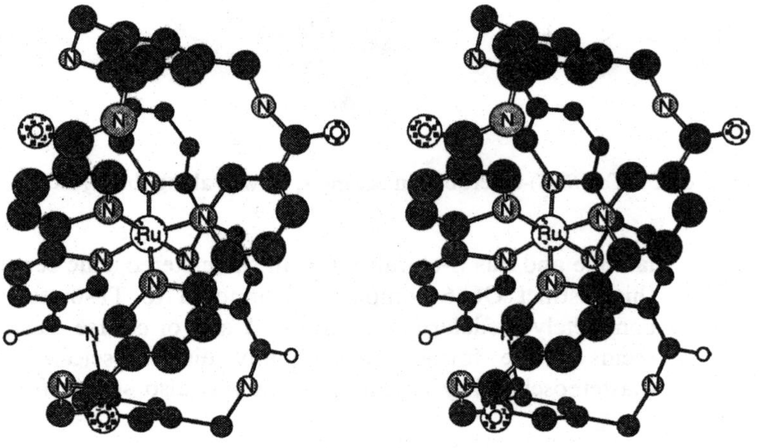

Figure 5.78
A tris(bpy)$_3$ cage [173]. (i) Schematic representation; (ii) stereo pair of the Δ-enantiomer

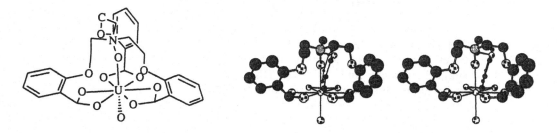

Figure 5.79
(i) Hexadentate ligand for the stereognostic complexation with uranyl ion. (ii) Stereo pair of the UO_2^{2+} complex

ligating atoms are arranged in a plane, with the two oxo ligands in apical positions. The complex stability is enhanced by a hydrogen bridge from the basal amine nitrogen to one of the oxo ligands. This example shows another possibility of modern synthetic chemistry, which allows for tailor-made ligands in those cases where a metal has specific non-spherical properties, such as the oxo cations. This approach is called *stereognostic coordination chemistry* [183]. It will enable chemists to built even more specific molecules for well defined purposes.

5.8 Ligands with Special Topologies

In a relatively recent development, coordination compounds of considerable complexity have been studied. The basis for this interest is on the one hand the realization that some important natural systems, such as photosystem II, use coordination units arranged in a highly organized manner [184], and on the other hand the perspective to construct artificial molecular devices, utilizing the versatile properties of metallic elements [162,185]. It became possible only recently, of course, to tackle the problem of analyzing these complicate systems in nature, and to construct the systems needed for molecular devices. The synthetic procedures have not changed drastically, but it is the methodology of modern analytical chemistry, especially NMR spectroscopy and X-ray diffraction, which has made this development possible.

Many of these more complicated systems involve polynuclear coordination compounds, where metals can carry out various functions, especially transfer of charge and energy in a highly organized manner. Some of these systems will be discussed in Chapter 6. Here mononuclear units are discussed, which are often at the basis of more complicated structures.

5.8.1 *Trans*-Spanning Ligands

Ligand *trans* positions are especially important in SP-4 and in OC-6 coordination geometries, although of course it is not restricted to these two. 'Ordinary' chelates occupy *cis* positions in these coordination geometries.

There were early attempts to synthesize *trans* complexes with bidentate ligands using long aliphatic chains between the ligating atoms. Schlesinger [186] in 1925 prepared copper(II) complexes of the ligand series (5.31). From the observation that the compounds are blue for $n = 2$ or 3, but violet for $n = 10$ and blue or violet for $n = 7$, he assumed that the long-chain ligands are *trans* spanning. Similarly, Issleib and Hohlfeld [187] obtained different compounds with the diphosphines shown in (5.32) as ligands in Ni^{II} complexes, only that the difference appears here for $n = 3$ and $n = 5$. Mochida *et al.* [188] attempted the *trans* coordination by using a terdentate amine in a Pt^{IV} complex, which was then transformed into a bidentate ligand spanning *trans* positions in an SP-4 Pt^{II} complex. This complex is, however, fairly unstable and reverts easily back to a normally bonded ligand.

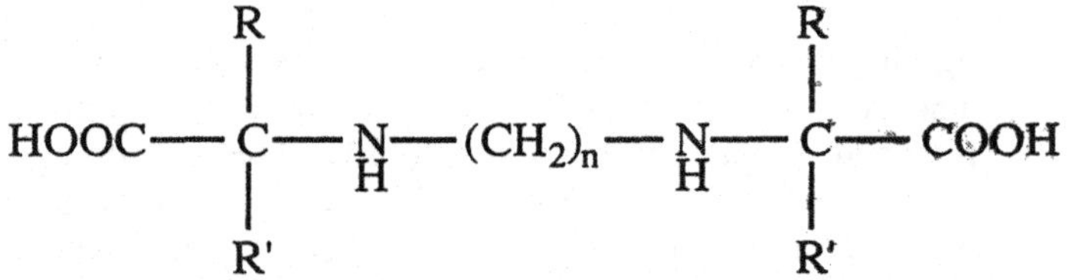

(5.31)

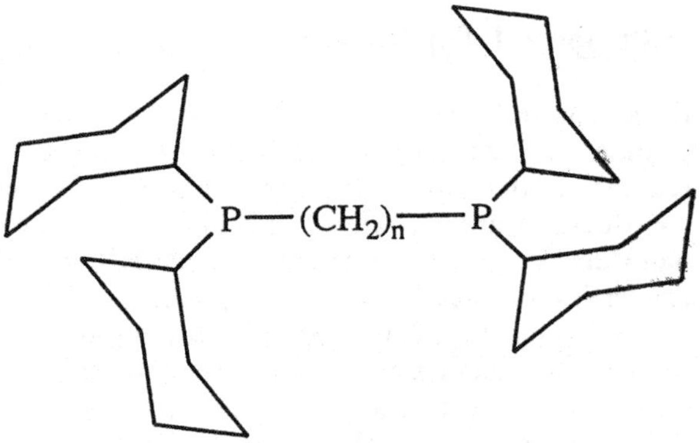

(5.32)

In a long series of publications, Venanzi and co-workers [189,190] reported many complexes with a rigid type of a bidentate phosphine ligand (Figure 5.80), which forms *trans* SP-4 complexes with Ni^{II}, Pd^{II}, and Pt^{II}. This is another example of a ligand in which the stereochemistry of the complex is predetermined by the ligand itself.

Trans spanning was also achieved for the octahedron, again with a bidentate diphosphine ligand [191]. In this case, the large chelate ring needed to span the *trans* positions was closed after two monodentate phosphine ligands were coordinated to the Ru^{II} central metal (Figure 5.81). The ring has a total of 21 atoms (1Ru, 2P, 2N, 16C).

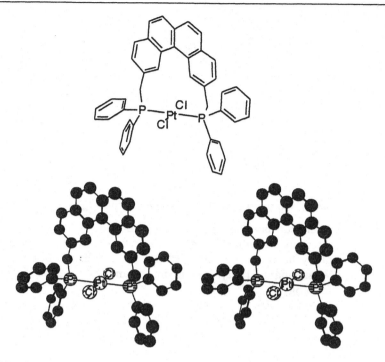

Figure 5.80
Trans spanning phosphine ligands: (i) 2,11-bis(diphenylphosphinomethyl)benzo[*c*]phenanthrene;
(ii) *trans*-[MX₂(i)]; M═Ni; X═Cl, Br, I, NCS; M═Pd; X═Cl, Br, I; M═Pt; X═Cl, I

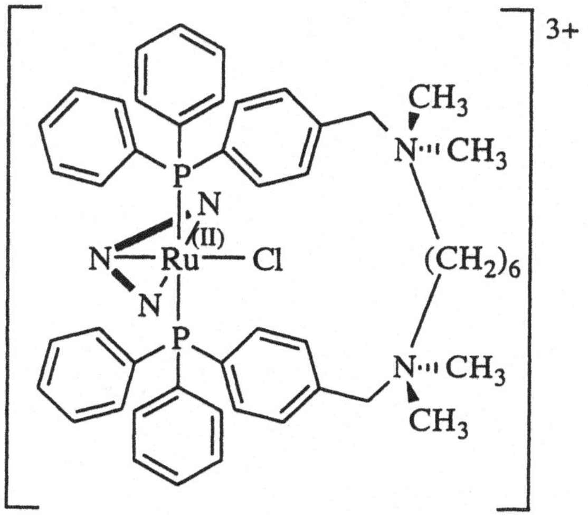

Figure 5.81
Trans OC-6 complex [191]. [(terpy)Ru^II(diphenyl-*P*-benzyl-N(Me)₂-(CH₂)₆-N(Me)₂-benzyl-*P*-
diphenyl)(Cl)](PF₆)₃ (where terpy═N—N—N═2,2′,2″-terpyridine)

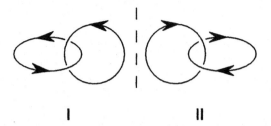

Figure 5.82
The two topological isomers of interconnected rings (Ref. 192, p. 3789).

5.8.2 Complexes with Interlocking Ring Ligands

Chemists have speculated about the possibility of the synthesis of two or more interlocking molecular rings for a long time. The two arrangements (Figure 5.82) of cyclic molecules (I) and (II) were called topological isomers by Frisch and Wassermann [192] in an important publication linking stereochemistry and topology.

A modern, efficient synthesis of two interlocking molecular rings has been developed by Dietrich-Buchecker and Sauvage [193], using a metal complex as an

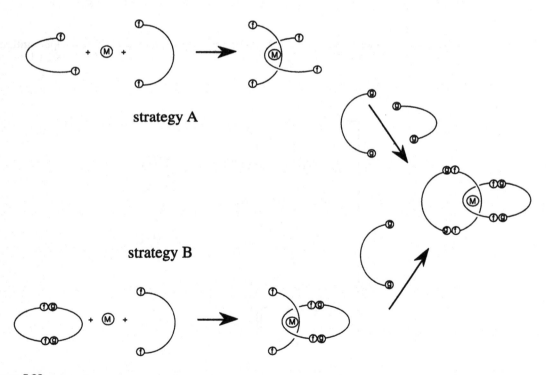

Figure 5.83
Strategies for the synthesis of interconnected rings using the template effect induced by a transition metal M. The groups f and g selectively bind to each other (Ref. 193, p. 801)

essential building tool. In this publication the historical development of interlocking rings in chemistry is also presented.

For the 'simple' [2]-catenand (two interlocked rings; higher catenands will be discussed in Chapter 6), two strategies were successfully developed (Figure 5.83). In both cases, the metal center is T-4 coordinated. The metal complex thus formed, called a catenate [193], can now be demetalated to form the desired structure of two interlocking rings. Only a transition metal can provide the combination of *thermodynamic stability* of the complex needed to permit the formation of the one (strategy A) or the two macrocyclic rings (strategy B) and the *kinetic lability* under certain conditions, which allows for the removal of the metal center. The syntheses have been realized with 2,9-substituted 1,10-phenanthroline and Cu^+ as the central metal. The two rings had 30 covalently bound atoms each (2N, 6O, 22C), and Cu^+ was eliminated by CN^- as a ligand. Although the dissociation of the copper catenate under the influence of cyanide is relatively slow, it is still fast enough for preparative purposes, showing again the general lability of T-4 metal coordination centers.

5.9   References

1. Brunner, H. (1969), *Angew. Chem., Int. Ed. Engl.*, **8**, 382–383.
2. Brunner, H. (1971), *Angew. Chem., Int. Ed. Engl.*, **10**, 249–260.
3. Brunner, H., *Transition Metal Chemistry and Optical Activity–Werner-Type Complexes, Organometallic Compounds, Enantioselective Catalysis*, in *Chirality–From Weak Bosons to the α-Helix*, R. Janoschek (Ed.), Springer, Berlin, 1991, pp. 166–179.
4. Brunner, H. and Schmidt, E. (1970), *J. Organomet. Chem.*, **21**, P53–P54.
5. Albright, T. A., Burdett, J. K. and Whangbo, M.-H., *Orbital Interactions in Chemistry*, Wiley, New York, 1985, p. 95.
6. Blackburn, B. K., Davies, S. G. and Whittaker, M., in *Stereochemistry of Organometallic and Inorganic Compounds*, I. Bernal (Ed.), Vol. 3, Elsevier, Amsterdam, 1989, Chapter 2.
7. Kauffman, G. B., *Classics in Coordination Chemistry, Classics of Science, Twentieth-Century Papers (1904–1935)*, Vol.8, Part 3, Dover, New York, 1978.
8. Kauffman, G. B., *Inorganic Coordination Compounds. Nobel Prize Topics in Chemistry, A Series of Historical Monographs on Fundamentals of Chemistry*, Heyden, London, 1981.
9. Vaska, L. and DiLuzio, J. W. (1961), *J. Am. Chem. Soc.*, **83**, 2784–2785.
10. Coussmaker, C. R. C., Hutchinson, M. H., Mellor, J. R., Sutton, L. E. and Venanzi, L. M. (1961), *J. Chem. Soc.*, **3**, 2705–2713.
11. Kilbourn, B. T., Powell, H. M. and Darbyshire, J. A. C. (1963), *Proc. Chem. Soc.*, 207–208.
12. King, R. B. and Reimann, R. H. (1976), *Inorg. Chem.*, **15**, 179–183.
13. Brunner, H. (1974), *Ann. N. Y. Acad. Sci.*, **239**, 182–192.
14. Brunner, H., *Top. Curr. Chem.*, **56**, 67–90.
15. Brunner, H. (1977), *Chem. Unserer Zeit*, **11**, 157–164.
16. Mestroni, G., Alessio, E., Zassinovich, G. and Marzilli, L. G. (1991), *Comments Inorg. Chem.*, **12**, 67–91.
17. Reisner, G. M., Bernal, I., Brunner, H., Muschiol, M. and Siebrecht, B. (1978), *J. Chem. Soc., Chem. Commun.*, **16**, 691–692.
18. Raymond, K. N., Corfield, P. W. R. and Ibers, J. A. (1968), *Inorg. Chem.*, **7**, 1362–1372.
19. Haaland, A., Hammel, A., Rypdal, K. and Volden, H. V. (1990), *J. Am. Chem. Soc.*, **112**, 4547–4549.

20. Leigh, G. J., *Nomenclature of Inorganic Chemistry*, Blackwell, Oxford, 1990.
21. Essen, L. N. and Gelman, A. D. (1956), *Zh. Neorg. Khim.*, **1**, 2475.
22. Essen, L. N., Zakharova, F. A. and Gelman, A. D. (1958), *Zh. Neorg. Khim.*, **3**, 2654–2661.
23. Bailar, J. C., Jr (1957), *J. Chem. Educ.*, **34**, 334–341.
24. Mayper, S. A. (1957), *J. Chem. Educ.*, **34**, 623.
25. Shimba, S., Fujinami, S. and Shibata, M. (1979), *Chem. Lett.*, 783–784.
26. Von Zelewsky, A. (1968), *Helv. Chim. Acta*, **51**, 803–807.
27. Kang, S. K., Tang, H. and Albright, T. A. (1993), *J. Am. Chem. Soc.*, **115**, 1971–1981.
28. Morse, P. M. and Girolami, G. S. (1989), *J. Am. Chem. Soc.*, **111**, 4114–4116.
29. Eliel, E. L. and Wilen, S. H., *Stereochemistry of Organic Compounds*, Wiley-Interscience, New York, 1994.
30. Jørgensen, C. K., *Inorganic Complexes*, 2nd edn, Academic Press, London, 1963, p. 22.
31. Bock, C. W., Kaufman, A. and Glusker, J. P. (1994), *Inorg. Chem.*, **33**, 419–427.
32. Corey, E. J. and Bailar, J. C., Jr (1959), *J. Am. Chem. Soc.*, **81**, 2620–2629.
33. Hawkins, C. J. and Palmer, J. A. (1982), *Coord. Chem. Rev.*, **44**, 160.
34. Rasmussen, K., *Potential Energy Functions in Conformational Analysis, Lecture Notes in Chemistry*, Vol. 37, Springer, Berlin, 1985.
35. Schwarzenbach, G. (1952), *Helv. Chim. Acta*, **35**, 2344–2359.
36. Schwarzenbach, G. (1973), *Chimia*, **27**, 115.
37. Mills, W. H. and Gotts, R. A. (1926), *J. Chem. Soc.*, 3121–3131.
38. Liu, J. C. I. and Bailar, J. C., Jr (1951), *J. Am. Chem. Soc.*, **73**, 5432–5433.
39. Bernal, I., LaPlaca, S., Korp, J., Brunner, H. and Herrmann, W. A. (1978), *Inorg. Chem.*, **17**, 382–388.
40. Kepert, D. L., *Inorganic Stereochemistry, Inorganic Chemistry Concepts*, Vol. 6, Springer, Berlin, 1982.
41. Bailar, J. C., Jr and Peppard, D. F. (1940), *J. Am. Chem. Soc.*, **62**, 105–109.
42. Bailar, J. C., Jr (1946), *Inorg. Synth.*, **2**, 222–225.
43. Werner, A. (1911), *Chem. Ber.*, **44**, 3272–3278.
44. Sievers, R. E., Moshier, R. W. and Morris, M. L. (1962), *Inorg. Chem.*, **1**, 966–967.
45. Cartwright, P., Gillard, R. D., Sillanpaa, R. and Valkonen, J. (1987), *Polyhedron*, **6**, 1775–1779.
46. Evans, E. H. M., Richards, J. P. G., Gillard, R. D. and Wimmer, F. L. (1986), *Nouv. J. Chim.*, **10**, 783–791.
47. Krause, R. A., Wickenden-Kozlowski, A. and Cronin, J. L. (1982), *Inorg. Synth.*, **21**, 12–16.
48. Gerlach, H. and Müllen, K. (1974), *Helv. Chim. Acta*, **57**, 2234–2237.
49. Eisenberg, R. and Ibers, J. A. (1966), *Inorg. Chem.*, **5**, 411–416.
50. Chassot, L., Mueller, E. and Von Zelewsky, A. (1984), *Inorg. Chem.*, **23**, 4249–4253.
51. Deuschel-Cornioley, C., Stoeckli-Evans, H. and Von Zelewsky, A. (1990), *J. Chem. Soc., Chem. Commun.*, 121–122.
52. Habu, T. and Bailar, J. C., Jr (1966), *J. Am. Chem. Soc.*, **88**, 1128–1130.
53. Raymond, K. N., Corfield, P. W. R. and Ibers, J. A. (1968), *Inorg. Chem.*, **7**, 842–844.
54. Raymond, K. N. and Duesler, E. N. (1971), *Inorg. Chem.*, **10**, 1486–1492.
55. Raymond, K. N. and Ibers, J. A. (1968), *Inorg. Chem.*, **7**, 2333–2338.
56. Hambley, T. W. (1987), *J. Comput. Chem.*, **8**, 651–657.
57. Jurnak, F. A. and Raymond, K. N. (1972), *Inorg. Chem.*, **11**, 3149–3152.
58. Niketic, S. R. and Woldbye, F. (1973), *Acta Chem. Scand.*, **27**, 621–642.
59. Nagao, R., Marumo, F. and Saito, Y. (1973), *Acta Crystallogr., Sect. B*, **29**, 2438–2443.
60. Jurnak, F. A. and Raymond, K. N. (1974), *Inorg. Chem.*, **13**, 2387–2397.
61. Thummel, R. P., Lefoulon, F. and Korp, J. D. (1987), *Inorg. Chem.*, **26**, 2370–2376.
62. Bernauer, K., *Diastereoisomerism and Diastereoselectivity in Metal Complexes*, in *Topics in Current Chemistry–Theor. Inorg. Chem. II*, Vol. 65, Springer, Berlin, 1976, pp. 1–35.
63. Borovik, A. S., Dewey, T. M. and Raymond, K. N. (1993), *Inorg. Chem.*, **32**, 413–421.
64. Branca, M., Micera, G., Dessi, A., Sanna, D. and Raymond, K. N. (1990), *Inorg. Chem.*, **29**, 1586–1589.

65. Brorson, M., Damhus, T. and Schäffer, C. E. (1983), *Inorg. Chem.*, **22**, 1569–1573.

66. Constable, E. C. (1986), *Adv. Inorg. Chem. Radiochem.*, **30**, 69–121.

67. Keene, F. R. and Searle, G. H. (1972), *Inorg. Chem.*, **11**, 148–156.

68. Keene, F. R., Searle, G. H., Yoshikawa, Y., Imai, A. and Yamasaki, K. (1970), *J. Chem. Soc., Chem. Commun.*, 784–786.

69. Kobayashi, M., Marumo, F. and Saito, Y. (1972), *Acta Crystallogr., Sect. B*, **28**, 470–474.

70. Geue, R. J. and Snow, M. R. (1977), *Inorg. Chem.*, **16**, 231–241.

71. Selbin, J. and Bailar, J. C., Jr (1960), *J. Am. Chem. Soc.*, **82**, 1524–1526.

72. Buckingham, D. A., Marzili, L. G. and Sargeson, A. M. (1967), *J. Am. Chem. Soc.*, **89**, 5133–5138.

73. Buckingham, D. A., Marzilli, L. G., Maxwell, I. E., Sargeson, A. M. and Freeman, H. C. (1969), *J. Chem. Soc. D*, **11**, 583–585.

74. Emmenegger, F. P. and Schwarzenbach, G. (1966), *Helv. Chim. Acta*, **49**, 625–643.

75. Muto, A., Marumo, F. and Saito, Y. (1970), *Acta Crystallogr., Sect. B*, **26**, 226–233.

76. Saito, Y., *Top. Stereochem.*, **10**, 95–174.

77. Sato, S. and Saito, Y. (1975), *Acta Crystallogr., Sect. B*, **31**, 2456–2460.

78. Yoshikawa, Y. (1976), *Bull. Chem. Soc. Jpn.*, **49**, 159–162.

79. Yoshikawa, Y. and Yamasaki, K. (1973), *Bull. Chem. Soc. Jpn.*, **46**, 3448–3452.

80. Okamoto, K., Tsukihara, T., Hidaka, J. and Shimura, Y. (1973), *Chem. Lett.*, 145–148.

81. Collins, J., Dwyer, F. P. and Lions, F. (1952), *J. Am. Chem. Soc.*, **74**, 3134–3136.

82. Dwyer, F. P. and Lions, F. (1947), *J. Am. Chem. Soc.*, **69**, 2917–2918.

83. Dwyer, F. P. and Lions, F. (1950), *J. Am. Chem. Soc.*, **72**, 1545–1550.

84. Dwyer, F. P., Lions, F. and Mellor, M. P. (1950), *J. Am. Chem. Soc.*, **72**, 5037–5039.

85. Bailar, J. C., Jr (1990), *Coord. Chem. Rev.*, **100**, 1–27.

86. Weatherall, D. J., Pippard, M. J. and Callender, S. T. (1983), *N. Engl. J. Med.*, **308**, 456–458.

87. Leong, J. and Raymond, K. N. (1975), *J. Am. Chem. Soc.*, **97**, 293–296.

88. Hossain, M. B., Jalal, M. A. F. and Van Der Helm, D. (1986), *Acta Crystallogr., Sect. C*, **42**, 1305–1310.

89. Bickel, H., Hall, G. E., Keller–Schierlein, W., Prelog, V., Vischer, E. and Wettstein, A. (1960), *Helv. Chim. Acta*, **43**, 2129–2138.

90. Matzanke, M., Berthold, F., Muller, G. I. and Raymond, K. N. (1984), *Biochem. Biophys. Res. Commun.*, **121**, 922–930.

91. Müller, G., Matzanke, B. F. and Raymond, K. N. (1984), *Biochem. Biophys. Res. Commun.*, **121**, 922–930.

92. Busch, D. H. and Bailar, J. C., Jr (1956), *J. Am. Chem. Soc.*, **78**, 716–719.

93. Harrowfield, J. M. and Wild, S. B., in *Comprehensive Coordination Chemistry*, Vol. 1, G. W. Wilkinson (Ed.), Pergamon Press, Oxford, 1987, pp.179–212.

94. Shimura, Y. (1958), *Bull. Chem. Soc. Jpn.*, **31**, 315–319.

95. Werner, A. (1918), *Helv. Chim. Acta*, **1**, 5–32.

96. Cooley, W. E., Liu, C. F. and Bailar, J. C., Jr (1959), *J. Am. Chem. Soc.*, **81**, 4189–4195.

97. Dunlop, J. H. and Gillard, R. D. (1965), *J. Chem. Soc.*, 6531–6541.

98. Mori, M., Shibata, M., Kyono, E. and Kanaya, M. (1961), *Bull. Chem. Soc. Jpn.*, **34**, 1837–1842.

99. Dunlop, J. H. and Gillard, R. D. (1966), *Adv. Inorg. Chem. Radiochem.*, **9**, 185–215.

100. Dunlop, J. H., Gillard, R. D., Payne, N. C. and Robertson, G. B. (1966), *J. Chem. Soc., Chem. Commun.*, **23**, 874–876.

101. Dunlop, J. H., Gillard, R. D. and Ugo, R. (1966), *J. Chem. Soc. A*, **11**, 1540–1547.

102. Gillard, R. D. (1967), *Inorg. Chim. Acta*, **1**, 69–86.

103. Harnung, S. E., Kallesoe, S., Sargeson, A. M. and Schaeffer, C. E. (1974), *Acta Chem. Scand., Ser. A*, **28**, 385–398.

104. Dwyer, F. P. and Garvan, F. L. (1959), *J. Am. Chem. Soc.*, **81**, 1043–1045.

105. Dwyer, F. P., Sargeson, A. M. and James, L. B. (1964), *J. Am. Chem. Soc.*, **86**, 590–592.

106. Kojima, M., Yoshikawa, Y. and Yamasaki, K. (1973), *Inorg. Nucl. Chem. Lett.*, **9**, 689–692.

107. Iwasaki, H. and Saito, Y. (1966), *Bull. Chem. Soc. Jpn.*, **39**, 92–100.

108. Kuroda, R. and Saito, Y. (1974), *Acta Crystallogr., Sect. B*, **30**, 2126–2130.
109. Harnung, S. E., Sørensen, B. S., Creaser, I., Maegaard, H., Pfenninger, U. and Schäffer, C. E. (1976), *Inorg. Chem.*, **15**, 2123–2126.
110. Kobayashi, A., Marumo, F. and Saito, Y. (1972), *Acta Crystallogr., Sect. B*, **28**, 2907–2915.
111. Saito, Y., *Inorganic Molecular Dissymmetry, Inorganic Chemistry Concepts*, Vol. 4, Springer, Berlin, 1979.
112. Abu-Dari, K. and Raymond, K. N. (1992), *J. Coord. Chem.*, **26**, 1–14.
113. Karpishin, T. B., Stack, T. D. P. and Raymond, K. N. (1993), *J. Am. Chem. Soc.*, **115**, 182–192.
114. Karpishin, T. B., Stack, T. D. P. and Raymond, K. N. (1993), *J. Am. Chem. Soc.*, **115**, 6115–6125.
115. Kobayashi, A., Marumo, F. and Saito, Y. (1974), *Acta Crystallogr., Sect. B*, **30**, 1495–1498.
116. Gollogly, J. R. and Hawkins, C. J. (1967), *Aust. J. Chem.*, **20**, 2395–2402.
117. Buckingham, D. A., Dwyer, F. P. and Sargeson, A. M., *Chelating Agents and Metal Chelates*, Academic Press, New York, 1964, p. 208.
118. Dwyer, F. P. and Garvan, F. L. (1961), *J. Am. Chem. Soc.*, **83**, 2610–2615.
119. Duncan, J. F. (1973), *Proc. R. Aust. Chem. Inst.*, **40**, 151–157.
120. Hahn, F. E., McMurry, T. J., Hugi, A. and Raymond, K. N. (1990), *J. Am. Chem. Soc.*, **112**, 1854–1860.
121. Hayoz, P. and Von Zelewsky, A. (1992), *Tetrahedron Lett.*, **33**, 5165–5168.
122. Hayoz, P., Von Zelewsky, A. and Stoeckli-Evans, H. (1993), *J. Am. Chem. Soc.*, **115**, 5111–5114.
123. Fenton, R. R., Stephens, F. S., Vagg, R. S. and Williams, P. A. (1992), *Inorg. Chim. Acta*, **197**, 233–242.
124. Bernauer, K. and Pousaz, P. (1984), *Helv. Chim. Acta*, **67**, 796–803.
125. Stoeckli-Evans, H., Brehm, L., Pousaz, P., Bernauer, K. and Bürgi, H.-B. (1985), *Helv. Chim. Acta*, **68**, 185–191.
126. Ashby, M. T., Govindan, G. N. and Grafton, A. K. (1994), *J. Am. Chem. Soc.*, **116**, 4801–4809.
127. Meisenheimer, J., Angermann, L. and Holstein, H. (1924), *Liebigs Ann. Chem.*, **438**, 261–278.
128. Blount, J. F., Freeman, H. C., Sargeson, A. M. and Turnbull, K. R. (1967), *J. Chem. Soc., Chem. Commun.*, **7**, 324–325.
129. Fujita, M., Yoshikawa, Y. and Yamatera, H. (1976), *Chem. Lett.*, 959–962.
130. Halpern, B., Sargeson, A. M. and Turnbull, K. R. (1966), *J. Am. Chem. Soc.*, **88**, 4630–4636.
131. Larsen, S., Watson, K. J., Sargeson, A. M. and Turnbull, K. R. (1968), *J. Chem. Soc., Chem. Commun.*, **15**, 847–849.
132. Kuebler, J. R., Jr and Bailar, J. C., Jr (1952), *J. Am. Chem. Soc.*, **74**, 3535–3538.
133. Buckingham, D. A., Marzilli, P. A., Sargeson, A. M., Mason, S. F. and Beddoe, P. G. (1967), *J. Chem. Soc., Chem. Commun.*, **9**, 433–435.
134. Heineke, D., Franklin, S. and Raymond, K. N. (1994), *Inorg. Chem.*, **33**, 2413–2421.
135. Albinati, A., Lianza, F., Berger, H., Pregosin, P. S., Rüegger, H. and Kunz, R. W. (1993), *Inorg. Chem.*, **32**, 478–486.
136. Mills, W. H. and Quibell, T. H. H. (1935), *J. Chem Soc.*, 839-846.
137. Boeyens, J. C. A. and Dobson, S. M., in *Stereochemistry of Organometallic and Inorganic Compounds. Stereochemical and Stereophysical Behaviour of Macrocycles*, Vol. 2, I. Bernal (Ed.), Elsevier, Amsterdam, 1987, pp. 2–102.
138. Buschmann, H.-J., in *Stereochemistry of Organometallic and Inorganic Compounds. Stereochemical and Stereophysical Behaviour of Macrocycles*, Vol. 2, I. Bernal (Ed.), Elsevier, Amsterdam, 1987, pp. 103–185.
139. Matthes, K. E. and Parker, D., in *Stereochemistry of Organometallic and Inorganic Compounds. Stereochemical and Stereophysical Behaviour of Macrocycles*, Vol. 2, I. Bernal (Ed.), Elsevier, Amsterdam, 1987, pp. 187–226.
140. De Diesbach, H. and Von der Weid, E. (1927), *Helv. Chim. Acta*, **10**, 886–888.

141. Price, R., in *Comprehensive Coordination Chemistry. The Synthesis, Reactions, Properties & Applications of Coordination Compounds. Applications*, Vol. 6, G. W. Wilkinson (Ed.), Pergamon Press, Oxford, 1987, pp. 87–91.
142. Crowfoot-Hodgkin, D. (1965), *Angew. Chem.*, **77**, 954–962.
143. Curtis, N. F. (1960), *J. Chem. Soc.*, 4409–4413.
144. Pedersen, C. J. (1967), *J. Am. Chem. Soc.*, **89**, 7017–7036.
145. Gokel, G., *Crown Ethers and Cryptands, Monographs in Supramolecular Chemistry Series*, J. F. Stoddart (Ed.), Royal Society of Chemistry, Cambridge, 1991.
146. Inoue, Y. and Gokel, G. W., (Eds), *Cation Binding by Macrocycles–Complexation of Cationic Species by Crown Ethers*, Marcel Dekker, New York, 1990.
147. Melson, G. A., in *Coordination Chemistry of Macrocyclic Compounds*, G. A. Melson (Ed.), Plenum Press, New York, 1979, pp. 1 and 17.
148. Zompa, L. J. and Margulis, T. N. (1978), *Inorg. Chim. Acta*, **28**, L157.
149. Wieghardt, K., Bossek, U., Chaudhuri, P., Herrmann, W., Menke, B. C. and Weiss, J. (1982), *Inorg. Chem.*, **21**, 4308–4314.
150. Setzer, W. N., Ogle, C. A., Wilson, G. S. and Glass, R. S. (1983), *Inorg. Chem.*, **22**, 266–271.
151. Wieghardt, K., Kueppers, H. J. and Weiss, J. (1985), *Inorg. Chem.*, **24**, 3067–3071.
152. Zompa, L. J. and Margulis, T. N. (1980), *Inorg. Chim. Acta*, **45**, L263–L264.
153. Ito, T., Ito, H. and Toriumi, K. (1981), *Chem. Lett.*, 1101–1104.
154. Mertes, K. B. (1978), *Inorg. Chem.*, **17**, 49–52.
155. Alcock, N. W., Herron, N. and Moore, P. (1978), *J. Chem. Soc., Dalton Trans.*, 394–399.
156. Louis, R., Metz, B. and Weiss, R. (1974), *Acta Crystallogr., Sect. B*, **30**, 774–780.
157. Louis, R., Agnus, Y. and Weiss, R. (1979), *Acta Crystallogr., Sect. B*, **35**, 2905–2910.
158. Bombieri, G., Forsellini, E., Del Pra, A., Cooksey, C. J., Humanes, M. and Tobe, M. L. (1982), *Inorg. Chim. Acta*, **61**, 43–49.
159. Searle, G. H. (1989), *Bull. Chem. Soc. Jpn.*, **62**, 4021–4032.
160. Yoshikawa, Y., Toriumi, K., Ito, T. and Yamatera, H. (1982), *Bull. Chem. Soc. Jpn.*, **55**, 1422–1424.
161. Stoddard, F. J. (1987), *Top. Stereochem.*, **17**, 207.
162. Lehn, J.-M. (1988), *Angew. Chem., Int. Ed. Engl.*, **27**, 89–112.
163. Boston, D. R. and Rose, N. J. (1968), *J. Am. Chem. Soc.*, **90**, 6859–6860.
164. Sargeson, A. M. (1979), *Chem. Br.*, **15**, 23–27.
165. Sargeson, A. M. (1991), *Chem. Aust.*, **58**, 176–178.
166. Comba, P. (1989), *Inorg. Chem.*, **28**, 426–431.
167. Comba, P., Sargeson, A. M., Engelhardt, L. M., Harrowfield, J. M., White, A. H., Horn, E. and Snow, M. R. (1985), *Inorg. Chem.*, **24**, 2325–2327.
168. Gahan, L. R., Hambley, T. W., Sargeson, A. M. and Snow, M. R. (1982), *Inorg. Chem.*, **21**, 2699–2706.
169. Garrett, T. M., McMurry, T. J., Hosseini, M. W., Reyes, Z. E., Hahn, F. E. and Raymond, K. N. (1991), *J. Am. Chem. Soc.*, **113**, 2965–2977.
170. McMurry, T. J., Hosseini, M. W., Garrett, T. M., Hahn, F. E., Reyes, Z. E. and Raymond, K. N. (1987), *J. Am. Chem. Soc.*, **109**, 7196–7198.
171. McMurry, T. J., Rodgers, S. J. and Raymond, K. N. (1987), *J. Am. Chem. Soc.*, **109**, 3451–3453.
172. Xu, J., Stack, T. D. P. and Raymond, K. N. (1992), *Inorg. Chem.*, **31**, 4903–4905.
173. De Cola, L., Barigelletti, F., Balzani, V., Belser, P., Von Zelewsky, A., Voegtle, F., Ebmeyer, F. and Grammenudi, S. (1988), *J. Am. Chem. Soc.*, **110**, 7210–7212.
174. Libman, J., Tor, Y. and Shanzer, A. (1987), *J. Am. Chem. Soc.*, **109**, 5880–5881.
175. Tor, Y., Libman, J., Shanzer, A., Felder, C. E. and Lifson, S. (1992), *J. Am. Chem. Soc.*, **114**, 6661–6671.
176. Tor, Y., Libman, J., Shanzer, A. and Lifson, S. (1987), *J. Am. Chem. Soc.*, **109**, 6517–6518.
177. Loomis, L. D. and Raymond, K. N. (1991), *Inorg. Chem.*, **30**, 906–911.
178. Isied, S. S., Kuo, G. and Raymond, K. N. (1976), *J. Am. Chem. Soc.*, **98**, 1763–1767.
179. Yakirevitch, P., Rochel, N., Albrecht-Gary, A.-M., Libman, J. and Shanzer, A. (1993), *Inorg. Chem.*, **32**, 1779–1787.

180. Belser, P., De Cola, L. and Von Zelewsky, A. (1988), *J. Chem. Soc., Chem. Commun.*, **15**, 1057–1058.
181. Rodriguez-Ubis, J.-C., Alpha, B., Plancherel, D. and Lehn, J.-M. (1984), *Helv. Chim. Acta*, **67**, 2264–2269.
182. Seel, C. and Voegtle, F., in *Perspectives in Coordination Chemistry*, C. Floriani, A. F. Williams and A. E. Merbach, (Eds), Verlag Helvetica Chimica Acta, Weinheim, 1992, pp. 31–53.
183. Franczyk, T. S., Czerwinski, K. R. and Raymond, K. N. (1992), *J. Am. Chem. Soc.*, **114**, 8138–8146.
184. Pascard, C., Guilhem, J., Chardon-Noblat, S. and Sauvage, J.-P. (1993), *New J. Chem.*, **17**, 331–335.
185. Balzani, V. and De Cola, L. (Eds), *Supramolecular Chemistry*, *NATO ASI Series*, Kluwer, Dordrecht, 1992.
186. Schlesinger, N. (1925), *Chem. Ber.*, 1877–1889.
187. Issleib, K. and Hohlfeld, G. (1961), *Z. Anorg. Allg. Chem.*, **312**, 169–179.
188. Mochida, I., Mattern, J. A. and Bailar, J. C., Jr (1975), *J. Am. Chem. Soc.*, **97**, 3021–3026.
189. Bürgi, H.-B., Murray-Rust, J., Camalli, M., Caruso, F. and Venanzi, L. M. (1989), *Helv. Chim. Acta*, **72**, 1293–1300.
190. De Stefano, N. J., Johnson, D. K. and Venanzi, L. M. (1974), *Angew. Chem., Int. Ed. Engl.*, **13**, 133–134.
191. Leising, R. A., Grzybowski, J. J. and Takeuchi, K. J. (1988), *Inorg. Chem.*, **27**, 1020–1025.
192. Frisch, H. L. and Wassermann, E. (1961), *J. Am. Chem. Soc.*, **83**, 3789–3795.
193. Dietrich-Buchecker, C. O. and Sauvage, J.-P. (1987), *Chem. Rev.*, **87**, 795–810.

6 Topographical Stereochemistry of Polynuclear Coordination Units

Chapters 1–5 are devoted to the consideration of essentially one coordination center. The great breadth of variability in coordination numbers and coordination geometries of metallic elements makes these considerations already considerably involved if one tries to be reasonably systematic and complete. The combination of several coordination centers in one molecular unit can therefore be extremely complex and an exhaustive consideration becomes an impossible task. There are two classes of multicenter metal compounds: those with direct metal to metal bonds and those where the centers are exclusively connected by bridging ligands. The first are generally called metal clusters. Here we treat only polynuclear complexes with bridging ligands, since the stereochemical principles discussed in Chapters 1–5 can be applied to such molecules, whereas clusters often need other types of stereochemical descriptors. Another class of polynuclear species which is not discussed here is that of oxo-bridged complexes, generally called poly-oxo-ions (Ref. 1, p.807). A wealth of information is available for these structurally highly interesting species, which occur mainly in compounds of the early transition elements.

A logical way of describing polynuclear coordination species is to select one of the metals as *the* coordination center, and to consider the other metals as being part of the ligands of the former. The selection of the coordination center may be natural in some cases or rather arbitrary, e.g. if the metallic centers are equivalent.

6.1 Polynuclear Complexes with Simple Bridging Ligands

We arbitrarily divide our discussion into two parts: (i) the polynuclear complexes with simple bridging ligands, and (ii) those with ligands that are designed to have a special influence on the stereochemistry of the whole molecule. Simple ligands are, e.g. hydride, the halides and pseudohalides, OH^-, and other small inorganic ligands, as well as simple organic molecules or ions. The special ligands, treated in Section 6.2, are those which are designed to give helices, higher catenates, knots, etc.

In the simplest case, two metals are connected by one or several bridges consisting of just one atom each. This is always the case for monoatomic bridging ligands such as the halides, thio, and oxo. Diatomic ligands can bridge two centers by one atom, e.g. OH^-, but also by two atoms, e.g. CN^-. The 'metal complex as ligand' concept is most easily applied in the case of monoatomic bridges, where *monoatomic* signifies the bridge itself, i.e. M—{μ-L(R′)}—M, and not necessarily the bridging ligand as a whole. The connection between two metal centers can be either through one bridge (a common corner of two coordination polyhedra), two bridges (a common edge), or three or more bridges (a common face). These three cases are schematically illustrated for the OC-6/OC-6 case in Figure 6.1.

The bridge in Figure 6.1(a) a can be either linear [2] or bent. The highest possible symmetries for an OC-6 monobridged dinuclear complex $(l_1)_5M\{\mu\text{-}(l_1)\}M(l_1)_5$ is C_{4v} for a linear and C_{2v} for a bent structure. A doubly bridged complex $(l_1)_4M\{\mu\text{-}(l_1)\}_2M(l_1)_4$ with coplanar M and bridging ligands (the most likely case for monoatomic ligands) also has C_{2v} symmetry, whereas the triply bridged $(l_1)_3M\{\mu\text{-}(l_1)\}_3M(l_1)_3$ has C_{3v} symmetry. All these structures are achiral.

Simple examples of trinuclear species with halide bridges have been described in gas-phase complexes, but also in the solid state. Various ways of linking polyhedra were described by Müller [3] in a book on inorganic structural chemistry.

Going from monoatomic to diatomic ligands, e.g. considering the bridging OH^-

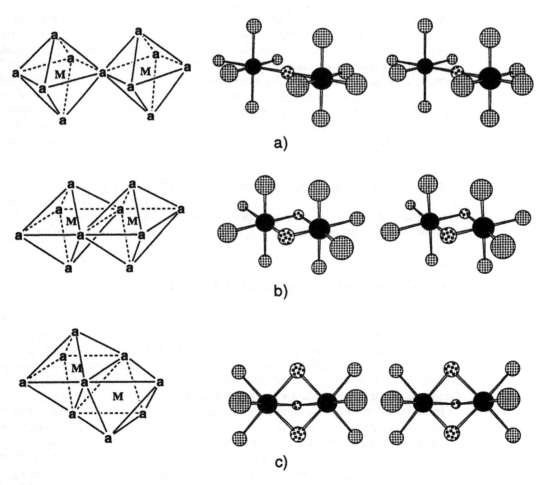

Figure 6.1
Representation of three different OC-6/OC-6 links with one, two, and three spherical bridging ligands

ligand, changes the symmetry of the species. As with many other diatomic ligands, OH^- is an ABE_3 species, having three lone pairs at the ligating atom. The VSEPR model tells us that the four atoms in the M—(OH)—M bridging unit will most likely not be coplanar, i.e. oxygen has a TPY-3 CG. A complex $(l_1)_5M\{\mu\text{-}OH\}M(l_1)_5$ will therefore have C_s symmetry [Figure 6.2(a)], but a heterometallic complex $(l_1)_5M\{\mu\text{-}OH\}M'(l_1)_5$ should be chiral, because the oxygen is stereogenic with three different ligand atoms (M,M',H) and one lone pair. The enantiomers thus obtained will, however, racemize rapidly.

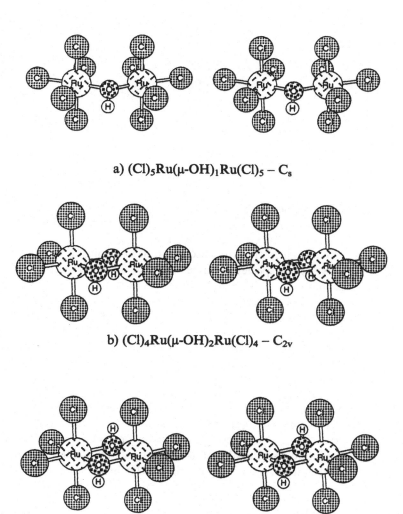

a) $(Cl)_5Ru(\mu\text{-}OH)_1Ru(Cl)_5 - C_s$

b) $(Cl)_4Ru(\mu\text{-}OH)_2Ru(Cl)_4 - C_{2v}$

b) $(Cl)_4Ru(\mu\text{-}OH)_2Ru(Cl)_4 - D_{2h}$

Figure 6.2
Diastereoisomers of complexes with OH^- bridges taking into account the lone pairs of the oxygen centers. (a) μ-OH, symmetry C_s; (b) $(\mu\text{-}OH)_2$, symmetry C_{2v}; (c) $(\mu\text{-}OH)_2$, symmetry D_{2h}; (d) $(\mu\text{-}OH)_3$, symmetry C_s; $(\mu\text{-}OH)_3$, symmetry D_{3h}

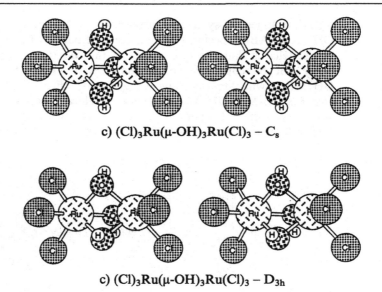

c) $(Cl)_3Ru(\mu\text{-}OH)_3Ru(Cl)_3 - C_s$

c) $(Cl)_3Ru(\mu\text{-}OH)_3Ru(Cl)_3 - D_{3h}$

Figure 6.2 *Continued*

The doubly bridged species $(l_1)_4M\{\mu\text{-}OH\}_2M(l_1)_4$ can exist in two diastereomeric forms, either with C_s [Figure 6.2 (b)] or D_{2h} [Figure 6.2 (c)] symmetry.[†] The triply bridged $(l_1)_3M\{\mu\text{-}OH\}_3M(l_1)_3$ has again two possibilities, having either C_s [Figure 6.2 (d)] or D_{3h} symmetry [Figure 6.2 (e)].

We shall restrict our discussion of examples of the topological stereochemistry of polynuclear species to some relatively simple cases, i.e. OC-6 and T-4 coordination geometries of the metal centers. A first and very elegant study on a polynuclear complex was carried out by Alfred Werner. In an attempt to show the applicability of the same stereochemical principles to organic and inorganic species, he was able to show [5] that the complex $[Co\{Co(NH_3)_4(OH)_2\}_3]^{6+}$ exists as two isolable enantiomers [6]. Figure 6.3 shows both forms of the complex, which can be imagined to be built up of a central cobalt(III) atom, coordinated by the three monopositive, bidentate cis-$Co^{III}(NH_3)_4(OH)_2$ 'ligands.'

It is therefore an OC-6 $[M(A^{\frown}A)_3]$ species, which is clearly chiral, having a maximum possible symmetry of D_3. The racemate was resolved by diastereomeric salt formation with the anion of bromocamphorsulfonic acid. Modern X-ray crystallographic investigations have completely confirmed the findings of Werner.[‡] An interesting detail, not yet imaginable by Werner, is the question of the planarity of the atoms in a 'chelate ring.' The bridging oxygen atoms become stereogenic

---

[†] Polynuclear complexes with bridges containing various ligands are discussed from a stereochemical point of view by Thewalt *et al.* [4]. They do not consider the non-planarity of the $M\{\mu\text{-}OH\}_2M$ skeleton, assigning to it an unconditional D_{2h} symmetry. The TPY-3 CG of oxygen in the bridging OH and its stereogenicity in bridges that are not symmetrical have in general not been considered, mainly owing to a lack of accurate experimental data. It should be taken into consideration for the discussion of more subtle stereochemical questions in polynuclear complexes.

[‡] Communicated by I. Bernal at the Coordination Chemistry Centenial Symposium (C3S), 205th National Meeting of the American Chemical Society, Denver, CO, 1993.

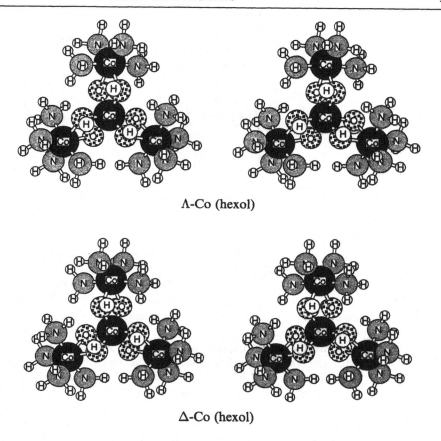

Λ-Co (hexol)

Δ-Co (hexol)

Figure 6.3
Stereo pairs of both enantiomers of the tetranuclear 'hexol' complex resolved by Werner (1914) [5]. The forms with homochiral bridging groups are depicted

upon coordination, since they have a TPY-3 coordination geometry with three different 'ligands' (plus a lone pair), which probably does not invert rapidly in the coordinated species. The X-ray structure determination shows that all OH groups in one octahedral face of the central cobalt ion are homochiral, either clockwise or anticlockwise oriented. In principle, the stereogenicity of the bridging oxygen centers causes a relatively large number of isomers. The configurations $\Delta(S)_6$, $\Delta(S)_5(R)$, $\Delta(S)_4(R)_2$, $\Delta(S)_3(R)_3$, etc., are all diastereomers. Those configurations with more than one oxygen of a given configuration have various possibilities of distributing the centers around the central metal, bringing the number of possible stereoisomers for the complex $[Co\{Co(NH_3)_4(OH)_2\}_3]^{6+}$ to 10 pairs of enantiomers (two D_3, one C_3, two C_2 and five C_1). These diastereomers will in general interconvert rapidly and the experiment shows that one form is more stable than the others, but one should be aware of this complication if, e.g., energy optimization calculations were carried out with such complexes. The 14 diastereomers will, most probably, represent local minima on an energy surface.

The tetranuclear complex $[Co\{Co(NH_3)_4(OH)_2\}_3]^{6+}$ is, from the stereochemical

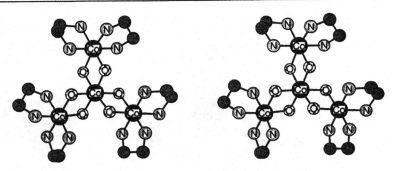

Figure 6.4
One isomer of the Co–en 'hexol' complex showing C_3 symmetry [7]

point of view, a relatively simple species compared with the chemically very similar $[Co\{Co(en)_2(OH)_2\}_3]^{6+}$ (Figure 6.4). In the latter all Co centers are helically chiral, resulting in four possible enantiomeric pairs $\{(\Delta(\Delta)_3/\Lambda(\Lambda)_3;\ \Delta\{(\Delta)_2\Lambda\}/\Lambda\{(\Lambda)_2\Delta;\ \Delta\{(\Lambda)_2\Delta\}/\Lambda\{(\Delta)_2\Lambda;\ \Delta(\Lambda)_3/\Lambda(\Delta)_3\}$, without considering the conformation of the en-chelates (δ/λ) of the peripheral metals, or the stereogenicity of the oxygen bridging centers.

If the δ/λ configurations are taken into account, 208 isomers are possible (2912 with the various configurations of the oxygen bridges) [4]. Several tetranuclear complexes of Co^{III} have been isolated and structurally characterized [4,8,9]. There exist isomers with mixed configurations such as $\Delta\{(\Lambda)_2\Delta\}/\Lambda\{(\Delta)_2\Lambda$ [9] and 'pure' configurations such as $\Delta(\Lambda)_3/\Lambda(\Delta)_3\}$ [7]. In the latter, the complex, which could have D_3 symmetry if all en rings would have the same conformation, exhibits only C_3 symmetry in the solid state, with alternating λ and δ conformations in the six en rings. The full stereochemical descriptor, as proposed by Thewalt *et al.* [4] (not taking into account the chirality at the oxygen centers), is therefore given by the following symbol (6.1):

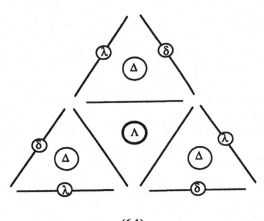

(6.1)

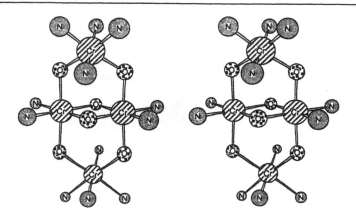

Figure 6.5
The chromium(III) 'hexol' skeleton. The symmetry is D_{2d} if the protons at the bridging OH groups are not taken into account

The configurations of the stereogenic oxygen centers cannot be discussed, since these crystal structure analyses gave no hydrogen positions.

The difficulty in predicting the stereochemistry of polynuclear complexes is clearly evidenced by another tetranuclear complex of the same general formula as Werner's hexol complexes, $[M_4(OH)_6(NH_3)_{12}]^{6+}$ and $[M_4(OH)_6(en)_6]^{6+}$. In the case of $M=Co^{III}$, the complexes discussed above seem to be the only products so far obtained, although they often represent a mixture of several closely related isomers. In the case of $M=Cr^{III}$, originally thought to have the same structures as the Co^{III} complexes [10], a completely different skeleton (Figure 6.5) was found by X-ray diffraction [11,12]. The Co—hexol skeleton seems to exist with Cr^{III} also [8].

The difference between the skeletons (the two type of structures are constitutional isomers) of the Co^{III} and the Cr^{III} complexes had been the subject of earlier conjecture on spectroscopic grounds [13]. An enumeration of the isomers of the Cr skeleton, taking into account the conformations of the en rings, yields 84 isomers for the complex with the en ligand [4]. The complex with ammonia ligands, on the other hand, has C_s symmetry, and it is therefore not chiral. Accordingly no enantiomers can occur, and no closely related diastereomers are expected either.

The complex $[Co\{Co(en)_2(OH)_2\}_3]^{6+}$, which exists in solution as a mixture of several of its 2912 stereoisomers, is another example of the *fuzzy stereochemistry* which can occur in metal complexes (see p. 132). This is especially pertinent in polynuclear complexes, where the metal centers are helically chiral. In many cases, polynuclear complexes have been prepared which are stereochemically not well defined, i.e. mixtures of many isomers will occur [14–16]. The tetranuclear complex $[Ru(bpym)_3(Ru(bpy)_2)_3]^{8+}$ (Figure 6.6), which has been studied spectroscopically [16], is the simplest case from a stereochemical point of view of a tetranuclear species with four helical chirality centers, because there are no other elements of chirality in the complex.

There are therefore 'only' eight isomers (four pairs of enantiomers), whose stereochemical symbols are the $\Delta(\Delta\Delta\Delta)/\Lambda(\Lambda\Lambda\Lambda)$, $\Delta(\Lambda\Delta\Delta)/\Lambda(\Delta\Lambda\Lambda)$, $\Delta(\Lambda\Delta\Delta)/\Lambda(\Delta\Lambda\Lambda)$, $\Delta(\Lambda\Lambda\Lambda)/\Lambda(\Delta\Delta\Delta)$. Undoubtedly, the spectroscopic and most other physical or chemical properties of these eight isomers are very similar, although no

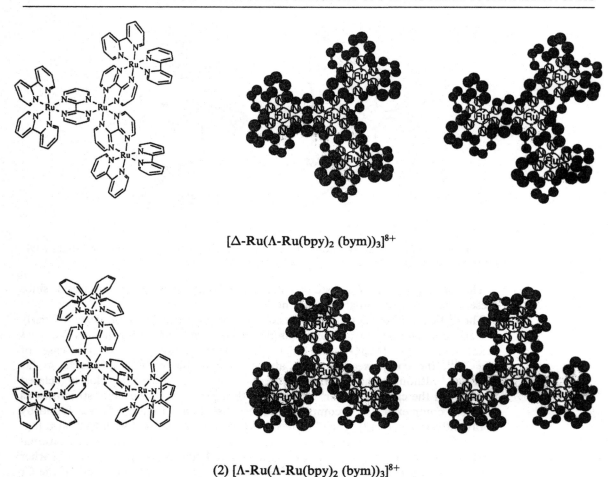

$$[\Delta\text{-}Ru(\Lambda\text{-}Ru(bpy)_2\ (bym))_3]^{8+}$$

$$(2)\ [\Lambda\text{-}Ru(\Lambda\text{-}Ru(bpy)_2\ (bym))_3]^{8+}$$

Figure 6.6
Two of the eight stereoisomers of the tetranuclear complex $[Ru(bpym)_3(Ru(bpy)_2)_3]^{8+}$. Stereopairs of the (a) $\Delta(\Lambda)_3$ and (b) $\Lambda(\Lambda)_3$ isomers are given

detailed investigation of an isomerically pure complex of this kind has yet been carried out. The complicated mixture, which cannot easily be separated into its components, prevents a complete characterization of the synthetically obtained material. In particular, the two most powerful methods for structural characterization, X-ray diffraction and NMR spectroscopy, cannot be applied: the former because no well defined crystalline material can be obtained,[†] and the latter because all diastereomers have different NMR spectra, rendering an assignment of NMR-absorptions almost impossible.

It is therefore important to develop a strategy for the synthesis of stereochemically well defined polynuclear species. The principle is clear, and it can be adopted from

[†]Even if in a preparation some crystalline material suitable for X-ray examination could be obtained, the result would be of limited value only, because the structure would not necessarily be significant for the bulk of the material (see also p. 16).

biochemistry, where larger molecules containing many chiral elements are almost always built up from enantiomerically pure building blocks (e.g. proteins, polysaccharides). Enantiomerically pure building blocks of metal complexes for polynuclear complexes of the type discussed above can be either configurationally stable, resolved (Δ/Λ) complexes of the type [cis-M(A^A)$_2$(l)$_2$], or complexes with a predetermined helicity at the metal center, like the chiragen complexes (see p. 139). Using Δ- and/or Λ-[Ru(bpy)$_2$(py)$_2$]$^{2+}$ or the corresponding phen complexes as building blocks, it was possible to obtain the complexes shown in Figure 6.7 in isomerically pure forms [17,18]. The same principle led to isomerically pure complexes with other bridging ligands [19,20].

Fuzzy stereochemistry can also be avoided in polypyridine metal complexes if, instead of bpy-type ligands, terpy units are used as building blocks [21]. Such complexes may, however, not have the interesting photophysical properties of the bpy-type compounds.

One of the prime interests in polynuclear complexes is their occurrence in a large number of biochemical systems. No element seems to form more polycentric complexes in nature than iron. Two principal types are known: oxo-bridged and sulfur-bridged polynuclear species. A review by Lippard [22] gives an account of several of the systems ranging from dinuclear species in hemerithin and in acid phosphatases to the high nuclearity of the ferritin core, which can store up to ca 4500 iron atoms in a spheroidal structure covered by a protein.

Iron has been known for a long time to form low molecular weight di- and trinuclear complexes. Two examples are dinuclear species bridged through a simple O(—II) oxo bridge, which seems to be linear [23,24].

Many examples of the so-called basic iron carboxylates, which had been known for a long time [25–27] show a triiron skeleton of D_{3h} symmetry with a central oxo ligand joining all three iron centers (Figure 6.8).

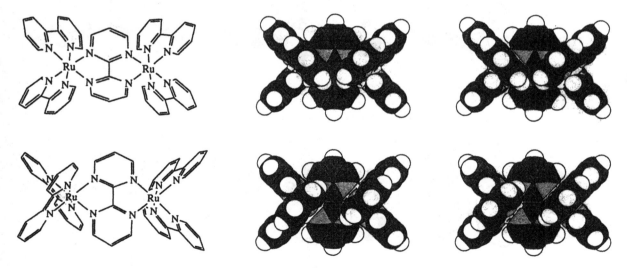

Figure 6.7
Stereo pairs of the two diastereoisomers (a) Δ,Δ[{Ru(bpy)$_2$}$_2$bpym]$^{4+}$ (C_2 symmetric, chiral), and (b) Δ,Λ[{Ru(bpy)$_2$}$_2$bpym]$^{4+}$ (C_s symmetric, achiral)

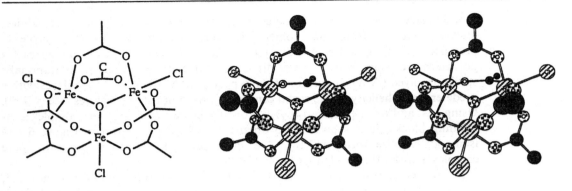

Figure 6.8
The D_{3h}-symmetric arrangement of the trinuclear basic iron carboxylates

Model complexes for the diiron centers, found in several biological systems, were prepared with two facially coordinating tridentate nitrogen donors, i.e. hydrotris(1-pyrazolyl)borate [28–30] (Figure 6.9) or with the small macrocycle 1,4,7-triazacyclononane [31] (Figure 6.10).

A cyclic polynuclear complex containing ten Zn centers has recently been characterized [32]. It has four T-4 ($ZnCl_4$), two T-4 [$Zn(H_2O)Cl_3$], two TB-5 ($ZnCl_2N_3$) and two distorted OC-6 [$Zn(OH_2)Cl_2N_3$] centers, showing the high flexibility of Zn as a coordination center.

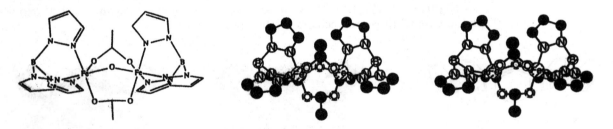

Figure 6.9
Model complexes for biologically important dinuclear iron complexes with 1,4,7-tricyclononane ligand [31]

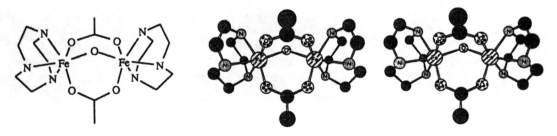

Figure 6.10
An analogous complex to that in Figure 6.9 with the hydrotris(1-pyrazolyl)borate [28,29]

6.2   Bridging Ligands for Molecular Helices, Chains, and Knots

The interlocking rings formed by the cyclic ligands in the Cu^I catenates, discussed in Chapter 5, Section 5.8.2, represent a special case of the general principle of predisposition of molecules (ligands) by complex formation for achieving a higher order structure or topology. In recent years, the application of this principle has given rise to various types of highly interesting structures. In mononuclear complexes, macrocycles, cages, and catenands were obtained in this way. Polynuclear systems offer even a wider variety of possibilities.

6.2.1   Helicates

One type of a higher order structure is a *helical* arrangement of ligand molecules. In mononuclear complexes, two-bladed propellers (p. 107) and especially three-bladed propellers are well known and they represent helical arrangements of ligands, which are chiral objects to which Δ/Λ descriptors can be assigned. In principle, there is no limitation to the number of blades in a propeller of a mononuclear complex. Thus mononuclear complexes of metals are often helical building blocks, and their combination to form complexes of higher nuclearity can lead to double-, triple-, and higher order helices. Since in such arrangements every metal is a center of chirality, the relationship between them is of great importance. There·are three simple cases, which can be imagined (Figure 6.11), as follows. (a) All centers in a given polynuclear (at least dinuclear) system are homochiral. In this case, synthesis from achiral or racemic building blocks will in general give a racemate (a racemic mixture, or a racemic modification) of the two types of polynuclear helices. In principle, this racemate can be separated into the enantiomeric forms. This case is to be expected if the formation of one coordination center induces a second neighboring coordination center with the same chirality. In this way the chirality propagates through the whole molecule. (b) A second possibility is that the arrangement of the ligands is such that coordination of the ligands on one center induces the formation of the neighboring center with opposite chirality. In a polynuclear complex with an even number of metal centers, this will lead to an achiral molecule. Complexes with an odd number of metal centers will still be chiral molecules. (c) If there is no preference for the helicity of a neighboring center, a disordered structure will be formed.

These cases (a)–(c) correspond to periodical (period lengths = one complex unit) or random arrangements. There are, of course, other periodicities possible. If synthetic chemistry moves ahead at a pace, as it did in the past decade, we might see helices where a long helical strand is disrupted at predetermined positions. Such structures may have interesting applications for molecular devices and in biochemistry [33]. However, such perspectives are today still in a purely speculative state. Several examples of simple helical arrangements have been described in the recent literature.

Complexes having a dinuclear helical molecular architecture have been reported with various ligands. In a relatively early investigation, it had been shown [34] that the octaethyl formylbiliverdinate ligand (Figure 6.12) forms an SPY-5 monomeric complex, in which the apex of the SPY-5 is occupied by a water ligand. On dehydration, a dinuclear complex is formed, where two ligands bridge the two metal centers, which become T-4, in a helical manner.

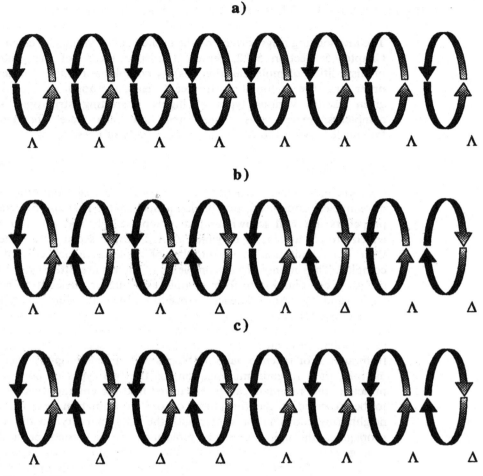

Figure 6.11
Three different arrangements of locally helical complexes in polynuclear species: (a) homochiral, forming a real molecular helix; (b) antichiral, forming species which are intramolecular racemates; (c) disordered

A second example of a dinuclear T-4 helix with a similar ligand [35,36] shows that the formation of a helix can be a consequence of a general structural principle.

The stereochemical influence of the metal undoubtedly plays an important role for the formation of such helical structures with the linear tetrapyrroles. An investigation of an Ni^{II} complex with a similar tetrapyrrole shows that Ni, which prefers an SP-4 coordination geometry with this type of ligand, forms a mononuclear species [37], despite the fact that there are strong steric repulsions between the oxygen ends of the ligand, forcing this complex to adopt a helical structure not unlike the two-bladed propeller discussed before (p. 167). Unlike the platinum complexes with two bidentate ligands, the structure of this mononuclear Ni complex represents a single-stranded helix (Figure 6.13).

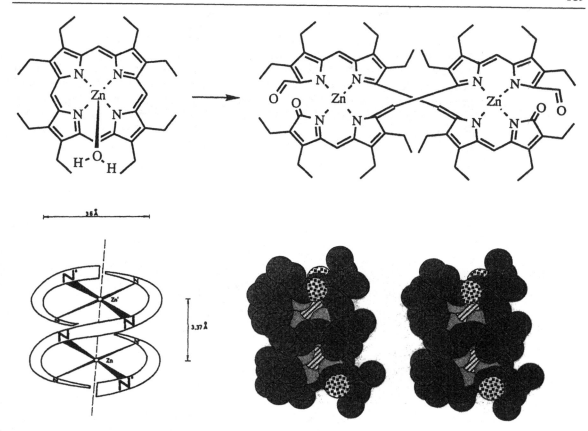

Figure 6.12
A homochiral dinuclear complex forming a double helix with local T-4 coordination (schematic after Ref. 34)

Figure 6.13
A single-stranded helix with local SP-4 coordination

The C_{2v}-symmetric tetradentate ligand (R,S)-1,2-(6-R'-py-2-CH=N)$_2$cy) (R'=H or CH$_3$, cy = cyclohexane) (Figure 6.14) was shown to form homochiral (Δ,Δ or Λ,Λ) helices with Ag^I and Cu^I as central metals [38,39].

The quinquepy ligand (Figure 6.15) forms with copper(II) acetate a dinuclear helical complex, with two differently coordinated Cu centers [40]. Two quinquepy ligands provide ten N ligand atoms. Together with a single coordinating acetate, one OC-6 (strongly distorted from O_h) CuN$_6$ and one TPY-5 CuN$_4$O form the two centers in the dimer.

Constable and co-workers [41–44] reported the synthesis of a series of further helical complexes with various metals.

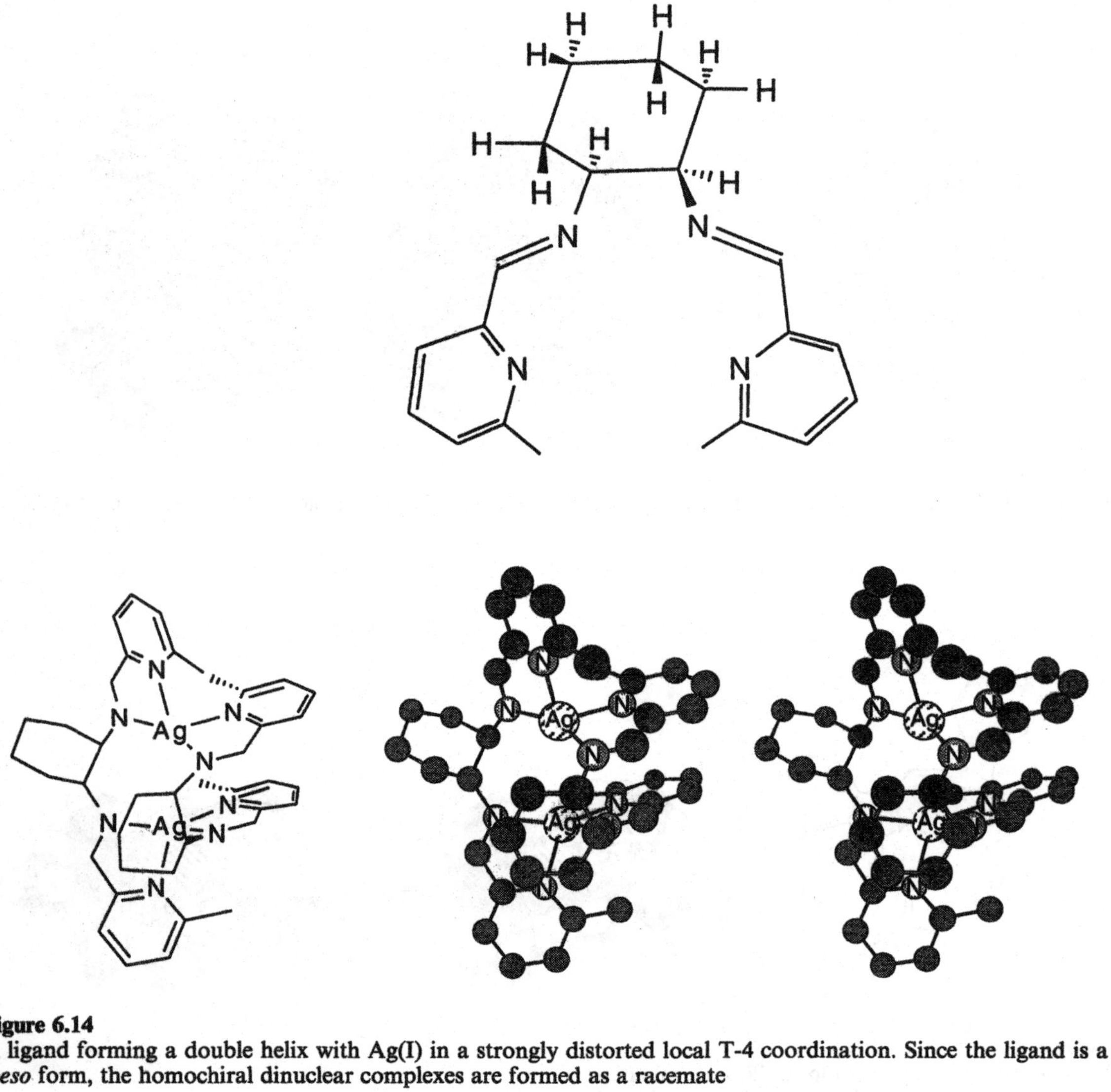

Figure 6.14
A ligand forming a double helix with Ag(I) in a strongly distorted local T-4 coordination. Since the ligand is a *meso* form, the homochiral dinuclear complexes are formed as a racemate

quinquepyridine

Figure 6.15
The highly irregular double helix formed by Cu(II) and the quinquepyridine ligand

Tri-, tetra-, and pentanuclear homochiral helicates have been obtained with Cu^I and bipyridine ligands connected through chains in 6,6′-positions (Figure 6.16) [45,46].

The stereochemistry of double helices formed by oligo(bipyridine) ligands in their Cu^I/Ag^I complexes has been discussed recently from the view point of its relation to the 'Coupe du Roi' [47]. La 'Coupe du Roi' is a cut of a sphere (e.g. an apple) yielding two homochiral halves. The helices so far synthesized with this type of ligands seem to form homochiral helices, which appear as racemates.

Triple helices require OC-6 building blocks. A dinuclear OC-6/OC-6 complex with three quadridentate ligands (M : L = 2 : 3) can again in principle give two arrangements: a homochiral and a heterochiral arrangement. The former has been described in a iron(II) complex with a flexible bridge between the bidentate moieties (6.2) [48]. The resulting structure can be considered to be triply helical, although the bridge between the two homochiral elements of the complex does not form a structure with an obviously helical structure.

This is the case in the Co^{II} complex described by Williams et al. [49]. The ligand is a relatively rigid bis(bidentate) molecule (6.3), where the second half of the molecule

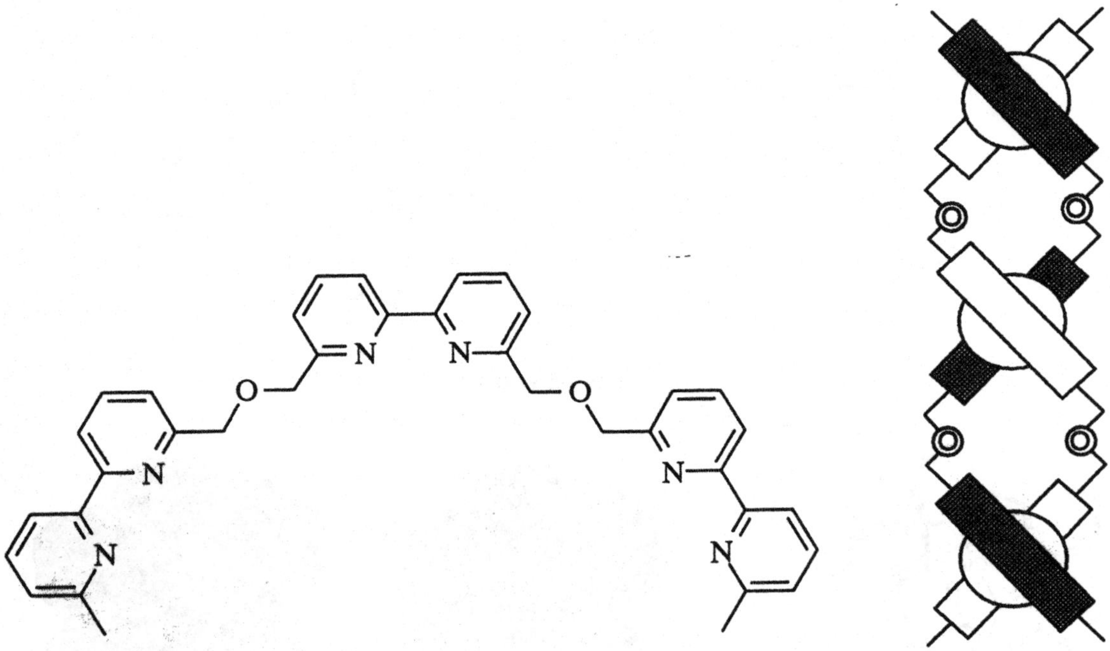

Figure 6.16
Schematic representation of the trinuclear triple helix formed by two ligands having each three bipyridine units, connected in 6,6′-positions. The coordinating units of one of the ligands are shaded

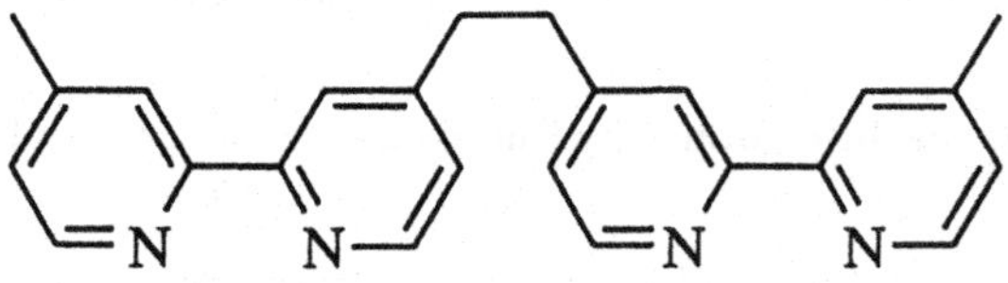

1,2-Bis[4-(4′-methyl = 2,2′-bipyridinyl)ethane

(6.2)

is predisposed to continue the helical structure created in the formation of the first coordination center. The symmetry (approximate in the solide state) of the complex is D_3, the same as that of an ordinary $[M(A^\frown A)_3]$ complex (Figure 6.17).

An early possible example of a triple helix in a dinuclear complex with three tetradentate ligands was reported by Stratton and Busch [50]. The ligand is pyridinaldizine (6.4). The various ways in which the ligand can coordinate to one or two metal centers led to the suggestion of designating it as 'flexidentate.' The question as to whether this dinuclear complex is homo- or heterochiral is not discussed in this publication.

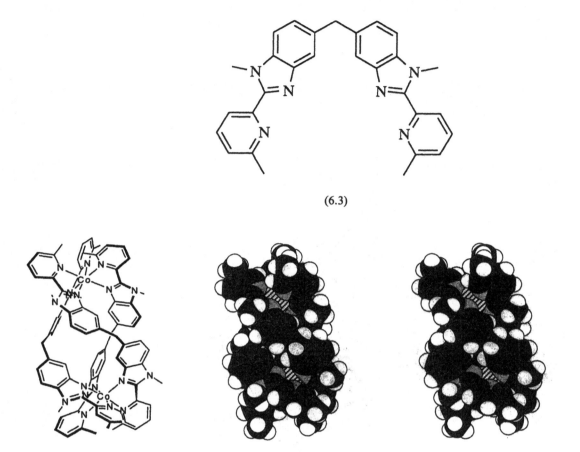

(6.3)

Figure 6.17
A dinuclear triple helix formed by a 3:2 complex of a tetradentate ligand with Co(II). The two metal sites are homotopic

(6.4)

A triply helical dinuclear iron(II) complex (Figure 6.18) with two non-identical sites has been described by Libman *et al.* [51].

The dinuclear iron(II) complex with bis(bipyridine) ligands bridged through the 5,5′-positions (6.5) [20] shows an interesting behavior. The ligand with $n = 1$ forms one single isomer of a 3:2 complex with Fe^{II}. This complex can be shown by NMR spectroscopy to be heterochiral. In a homochiral, helical dinuclear complex,

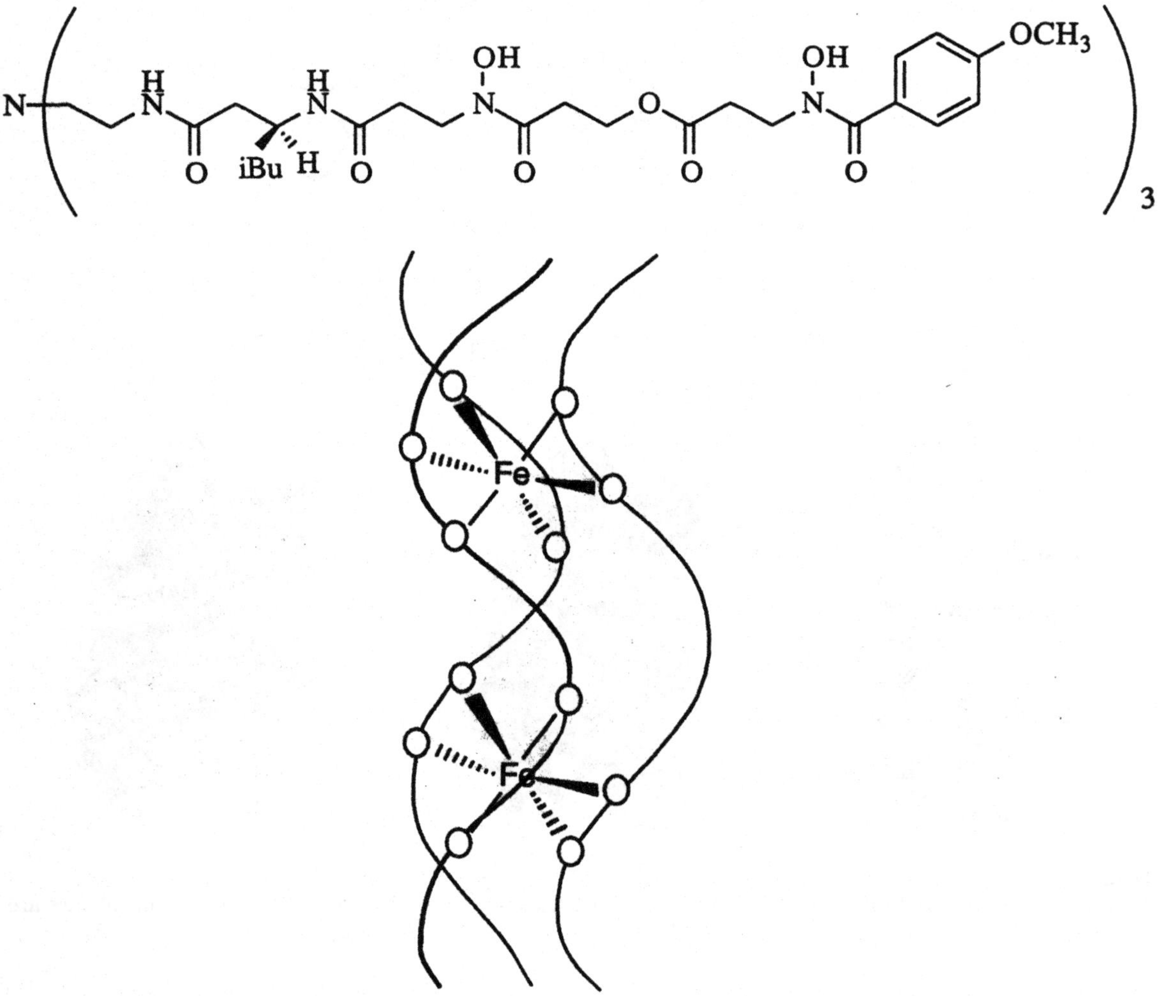

Figure 6.18
A dinuclear triple helix with two heterotopic metal sites

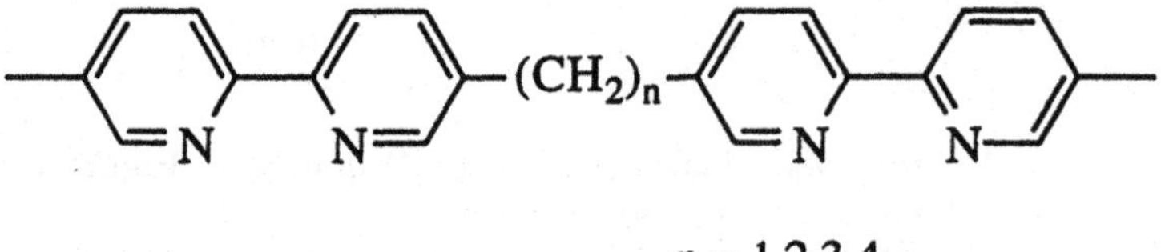

$$n = 1, 2, 3, 4$$

(6.5)

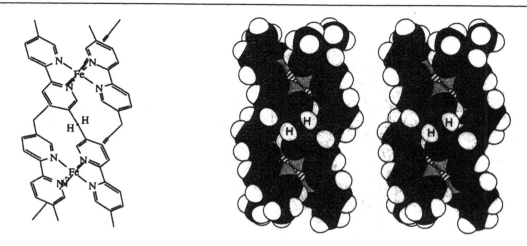

Λ,Λ-Fe$_2$(bpy[Cl]bpy)$_3$: helical arrangement

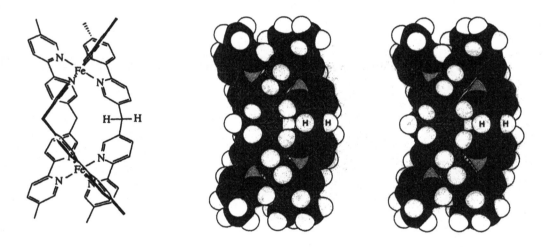

Δ,Λ-Fe$_2$(bpy[Cl]bpy)$_3$: antihelical arrangement

Figure 6.19
Stereo models of the homo- and heterochiral (antichiral) dinuclear complexes of Fe(II) with a bisbipyridine ligand (6.14, $n = 1$). The homochiral species has D_3 symmetry and the two protons of the CH$_2$ group are symmetry equivalent. The experimentally determined complex is C_{3h} symmetric and the two CH$_2$ protons are equivalent

the two protons of the CH$_2$ bridge would be symmetry equivalent through a C_2 axis (Figure 6.19), whereas in the heterochiral case the complex has C_{3h} symmetry and the two CH$_2$ protons lie on a mirror plane, but they are inequivalent. The AB system observed for these protons shows unambiguously that the latter type of

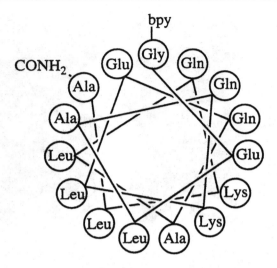

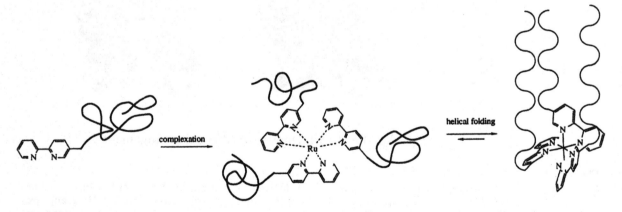

NH-Gly-Glu-Leu-Ala-Lys-Leu-Glu-Gln-Ala-Leu-Gln-Lys-Leu-Ala-NH$_2$

(6.6)

Figure 6.20
A triple helix of peptide chains, induced by an Ru(II) OC-6 complex

complex is formed. With longer bridges, both isomers (the homo- and the heterochiral) seem to be formed.

Triple helices formed by proteins have been obtained by assembling bipyridine derivatives of peptides with metal complexes (6.6) [52]. A Ru^{II} coordinated with

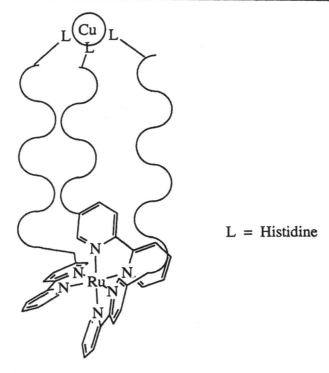

L = Histidine

Figure 6.21
A hetero-bimetallic complex, where the two metal centers are joined by a petide triple helix

three bpy-type ligands induces a triple helix, if the ligands coordinate in a facial way. The peptide chains then assemble through multiple weak interactions (Figure 6.20). If the peptide carries ligating groups in terminal positions, a second metal can coordinate at the opposite end of the triple helix (Figure 6.21) [53].

6.2.2 Molecular Chains and Knots

In Chapter 5, Section 5.8.2, complexes with interlocking ring ligands were presented. These mononuclear complexes are the smallest representatives of catenates, since a chain consists of minimally two links. The extension of the principles discussed in Section 5.8.2 to polynuclear complexes led Sauvage's group to the synthesis of a molecular chain with three links and a trefoil knot.

The synthetic principles are always based on a template reaction of a molecule predisposed in a structure which yields the desired topology. Thus, the three-link chain was synthesized adopting strategy B (p. 170), continuing with the coupling of two intermediates. The sequence of steps is depicted schematically in (6.7).

As a metal, again Cu^I with T-4 coordination geometry fulfils the job as a

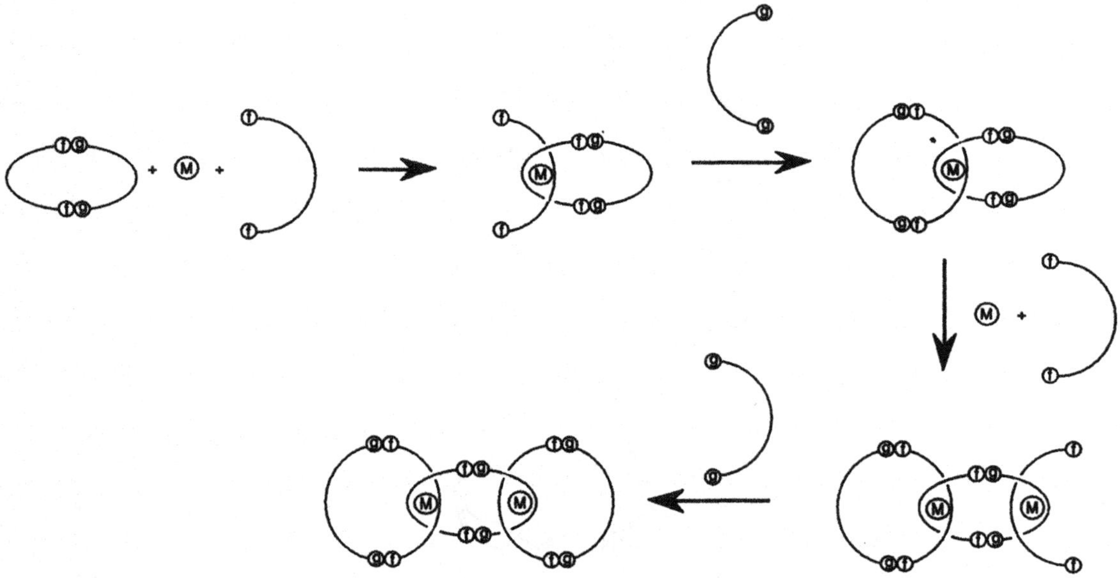

(6.7)

coordination center, forming complexes stable enough to perform organic reactions producing the rings, but sufficiently labile (and also thermodynamically stable) to be removed under mild conditions (always using CN^- as ligand), once the preorientation is no longer needed. An outline of the synthesis is shown in (6.8). If a similar principle is applied to a dinuclear helicate, even a molecular knot can be formed (6.9).

The structure of the dicopper(I) complex of the first knot of this type was confirmed by an X-ray structure in 1990 [54]. The tetradentate ligand, which was obtained after treatment of the complex with cyanide ions, is a knotted macrocycle of 86 covalently bound atoms (8 N, 14 O, 64 C). Its structure can be derived topologically (and chemically!) from a double helix, and it is consequently an intrinsically chiral object[†] whether it is present as the metal complex or as the free ligand. This was confirmed by the observation of a specific shifting, using Pirkle's reagent [55,56], which is often used to demonstrate chirality in a molecular system. Pirkle's reagent is an enantiomerically pure compound (6.10) that shifts NMR signals of enantiomers differently through stereospecific intermolecular interactions. Since the synthesis of the first knot was prepared, a number of variations of the original molecules have been published.

[†]Unlike many other chiral objects (e.g. a complex OC-6[M(A^A)₃], which can be transformed by a continuous reduction of the twist angle θ to 0° (p. 106) into an achiral TP-6 geometry), a molecular knot cannot be brought into an achiral form unless at least one covalent bond is broken. A *knot* can therefore be called a *topologically chiral object*.

(6.8)

(6.9)

$$F_3C \overset{OH}{\underset{H}{\cdots}}$$

(6.10)

6.3   References

1. Cotton, F. A. and Wilkinson, G., *Advanced Inorganic Chemistry*, 5th edn, Wiley, New York, 1988.
2. Wieghardt, K., Chaudhuri, P., Nuber, B. and Weiss, J. (1982), *Inorg. Chem.*, **21**, 3086–3090.
3. Müller, U., *Inorganic Structural Chemistry, Inorganic Chemistry: A Textbook Series*, Wiley, Chichester, 1994.
4. Thewalt, U., Jensen, K. A. and Schäffer, C. E. (1972), *Inorg. Chem.*, **11**, 2129–2136.
5. Werner, A. (1914), *Chem. Ber.*, **47**, 3087–3094.
6. Shimura, Y. (1984), *Rev. Inorg. Chem.*, **6**, 149–193.
7. Thewalt, U. and Ernst, J. (1975), *Z. Naturforsch., Teil B*, **30**, 818–819.
8. Andersen, P. and Berg, T. (1974), *J. Chem. Soc., Chem. Commun.*, 600–601.
9. Thewalt, U. (1971), *Chem. Ber.*, **104**, 2657–2669.
10. Pfeiffer, P., Voster, W. and Stern, R. (1908), *Z. Anorg. Chem.*, **58**, 272–296.
11. Bang, E. (1968), *Acta Chem. Scand.*, **22**, 2671–2684.
12. Flood, M. T., Marsh, R. E. and Gray, H. B. (1969), *J. Am. Chem. Soc.*, **91**, 193–194.
13. Bjerrum, J. (1964), *Quad. Chim., Cons. Naz. Rec. (Italy)*, **1**, p. 47.
14. Belser, P., Von Zelewsky, A., Frank, M., Seel, C., Voegtle, F., De Cola, L., Barigelletti, F. and Balzani, V. (1993), *J. Am. Chem. Soc.*, **115**, 4076–4086.
15. Denti, G., Serroni, C., Campagna, S., Juris, A., Ciano, M. and Balzani, V., in *Perspectives in Coordination Chemistry*, A. F. Williams, C. Floriani and A. E. Merbach (Eds), Verlag Helvetica Chimica Acta, Basle, 1992, pp. 153–164.
16. Hunziker, M. and Ludi, A. (1977), *J. Am. Chem. Soc.*, **99**, 7370–7371.
17. Hua, X., *Chiral Building Blocks Ru(L^L)₂ for Coordination Compounds*, Diss. No. 1047, University of Fribourg, Fribourg, 1993.
18. Hua, X. and Von Zelewsky, A. (1995), *Inorg. Chem.*, **34**, 5791–5794.
19. Jandrasics, E., *Synthesen chiraler Ru(II)- und Os(II)- Diimin-Komplexe unter Anwendung eines Mikrowellenofens*, Diss. No. 1085, University of Fribourg, Fribourg, 1995.
20. Nachbaur, J. A., *Metallkomplexe einer neuen Ligandfamilie mit zwei Bipyridinfunktionen*, Diss. No. 1042, University of Fribourg, Fribourg, 1993.
21. Newkome, G. R., Cardullo, F., Constable, E. C., Moorefield, C. N. and Thompson, A. M. W. C. (1993), *J. Chem. Soc., Chem. Commun.*, 925–927.
22. Lippard, S. J. (1988), *Angew. Chem., Int. Ed. Engl.*, **27**, 344–361.
23. Murray, K. S. (1974), *Coord. Chem. Rev.*, **12**, 1–35.
24. Schugar, H. J., Rossman, G. R., Barraclough, C. G. and Gray, H. B. (1972), *J. Am. Chem. Soc.*, **94**, 2683–2690.
25. Straugham, B. P. and Lam, O. M. (1985), *Inorg. Chim. Acta*, **98**, 7–10.
26. Thich, J. A., Toby, B. H., Powers, D. A., Potenza, J. A. and Schugar, H. J. (1981), *Inorg. Chem.*, **20**, 3314–3317.
27. Weinland, R., *Einführung in die Chemie der Komplexverbindungen*, Ferdinand Enke Verlag, Stuttgart, 1919, pp. 345*ff*, and references cited therein.
28. Armstrong, W. H. and Lippard, S. J. (1983), *J. Am. Chem. Soc.*, **105**, 4837–4838.
29. Armstrong, W. H., Spool, A., Papaefthymiou, C., Frankel, R. B. and Lippard, S. J. (1984), *J. Am. Chem. Soc.*, **106**, 3653–3667.

30. Trofimenko, S. (1970), *Inorg. Synth.*, **12**, 99–109.
31. Wieghardt, K., Pohl, K. and Gebert, W. (1983), *Angew. Chem., Int. Ed. Engl.*, **22**, 727.
32. Graf, M. and Stoeckli-Evans, H. (1994), *Acta Crystallogr.*, **C50**, 1461–1464.
33. Rehmann, J. P. and Barton, J. K. (1990), *Biochemistry*, **29**, 1701–1709.
34. Struckmeier, G., Thewalt, U. and Furhop, J. H. (1976), *J. Am. Chem. Soc.*, **98**, 278–279.
35. Sheldrick, W. S. and Engel, J. (1980), *J. Chem. Soc., Chem. Commun.*, 5–6.
36. Sheldrick, W. S. and Engel, J. (1981), *Acta Crystallogr., Sect. B*, **37**, 250–252.
37. Bonfiglio, J. V., Bonnet, R., Buckley, D. G., Hamzetash, D., Hursthouse, M. B., Malik, K. M. A., McDonagh, A. F. and Trotter, J. (1983), *Tetrahedron*, **39**, 1865–1871.
38. Van Stein, G. C., Van der Poel, H., Van Koten, G., Spek, A. L., Duisenberg, A. J. M. and Pregosin, P. S. (1980), *J. Chem. Soc., Chem. Commun.*, 1016–1018.
39. Van Stein, G. C., Van Koten, G., Vrieze, K., Brévard, C. and Spek, A. L. (1984), *J. Am. Chem. Soc.*, **106**, 4486–4492.
40. Constable, E. C., Drew, M. G. B. and Ward, M. D. (1987), *J. Chem. Soc., Chem. Commun.*, 1600–1601.
41. Constable, E. C. (1990), *Nature (London)*, **346**, 314–315.
42. Constable, E. C. (1992), *Tetrahedron*, **48**, 10013–10059.
43. Constable, E. C., Elder, S. M., Raithby, P. R. and Ward, M. D. (1991), *Polyhedron*, **10**, 1395–1400.
44. Constable, E. C. and Ward, M. D. (1990), *J. Am. Chem. Soc.*, **112**, 1256–1259.
45. Lehn, J.-M. and Rigault, A. (1988), *Angew. Chem., Int. Ed. Engl.*, **27**, 1095–1097.
46. Lehn, J.-M., Rigault, A., Siegel, J., Harrowfield, J., Chevrier, B. and Moras, D. (1987), *Proc. Natl. Acad. Sci. USA*, **84**, 2565–2569.
47. Glaser, R. (1993), *Chirality*, **5**, 272–276.
48. Serr, B. R., Andersen, K. A., Elliot, C. M. and Anderson, O. P. (1988), *Inorg. Chem.*, **27**, 4499–4504.
49. Williams, A. F., Piguet, C. and Bernadinelli, G. (1991), *Angew. Chem., Int. Ed. Engl.*, **30**, 1490.
50. Stratton, W. J. and Busch, D. H. (1958), *J. Am. Chem. Soc.*, **80**, 3191–3195.
51. Libman, J., Tor, Y. and Shanzer, A. (1987), *J. Am. Chem. Soc.*, **109**, 5880–5881.
52. Ghadiri, M. R., Soares, C. and Choi, C. (1992), *J. Am. Chem. Soc.*, **114**, 825–831.
53. Ghadiri, M. R. and Case, M. A. (1993), *Angew. Chem., Int. Ed. Engl.*, **32**, 1594.
54. Dietrich-Buchecker, C. O., Guilhem, J., Pascard, C. and Sauvage, J.-P. (1990), *Angew. Chem., Int. Ed. Engl.*, **29**, 1154–1156.
55. Pirkle, W. and Hoekstra, M. S. (1976), *J. Am. Chem. Soc.*, **98**, 1832–1839.
56. Pirkle, W. H. and Hoover, D. J. (1982), *Top. Stereochem.*, **13**, 263–331.

7 The Stereochemical Course of Reactions of Metal Complexes

This chapter deals with the relationships of the topographical stereochemical properties of reactants and products in chemical reactions where metal complexes participate. It is not intended to give a detailed account of the mechanisms of such reactions, although in a few cases some mechanistic viewpoints will be mentioned. For a discussion of mechanistic aspects in connection with the stereochemical course of reactions in coordination chemistry, the reader is referred to the book (especially Chapter 7) by Wilkins [1]. We shall follow a rather descriptive path, trying to present the most pertinent facts and some basic rules. It must be stated that even in 1995, a semi-quantitative theoretical understanding based on orbital interactions, such as that which has so much contributed to the progress in stereochemistry of purely organic compounds, is still in its infancy for most of coordination chemistry. No attempt is made, therefore, to base our considerations on a general theory of orbital interactions.

7.1 Isomerization and Substitution Reactions

The two types of reactions, isomerization and substitution, are closely related in many cases and they will be treated together. We shall restrict our discussion to SP-4 and OC-6 complexes, with a few remarks on some special T-4 and five-coordinated species, since most of our knowledge about topographical stereochemistry stems from complexes of these coordination geometries.

7.1.1 T-4 to SP-4 Polytopal Isomerization

As mentioned earlier, Ni^{II} complexes are the most likely to occur with similar energies in T-4 and in SP-4 geometries. A thoroughly investigated series of such molecules are the complexes with a bis(aminotroponeimato) (7.1) complex [2].

Since the spin state of the complexes changes during the isomerization T-4 ($S = 1$) to SP-4 ($S = 0$), NMR spectroscopy is ideally suited for the study of the process. These and similar isomerization processes have been reviewed in a paper which deals generally with the stereochemistry of bis(chelate)metal(II) complexes [3]. Depending strongly on the bulkiness of the R groups, the mole fraction of the tetrahedral form in thermodynamic equilibrium changes from ~ 0 (R = H) to 0.98 (R = n-Pr), the tetrahedral species being favored by the entropic term (higher degeneracy), but disfavored by the enthalpic term.

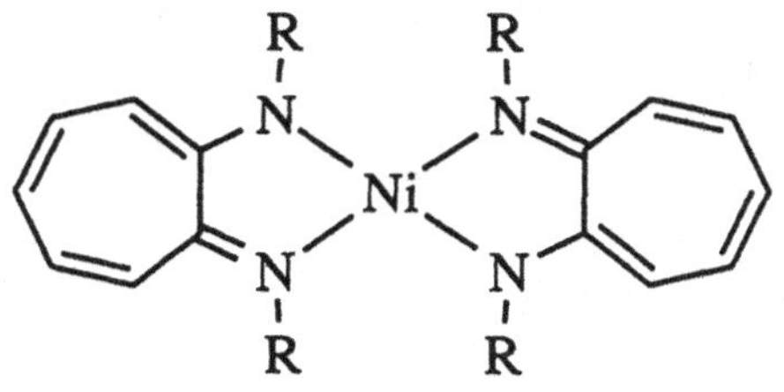

(7.1)

7.1.2 Substitution Reactions in SP-4 Complexes

The earliest systematic investigations on the stereochemical course of any kind of reactions in coordination chemistry were those concerning substitution reactions of SP-4 Pt^{II} complexes by Chernyaev [4]. In a successful attempt to synthesize all three diastereomers of a SP-4 [Mabcd] complex (as opposed to the two enantiomers of a corresponding T-4 complex), Chernyaev carried out a large number of substitution reactions with Pt^{II} complexes. He observed a striking regularity in the influence of a coordinated ligand to the substitution of the ligand in a *trans* position. The original publications are in Russian journals, which are difficult to obtain. The reader is referred to a paper by Kauffman [5] for a presentation of the historic details.

Chernyaev presented a series of ligands (extended later) which exert an increasing influence of directing an incoming ligand into *trans* positions. This is the so-called *trans*-effect series (7.2).

$$CN^- = CO = NO = H^- > CH_3^- = SC(NH_2)_2 = PR_3 > SO_3H^- >$$

$$NO_2^- = I^- = SCN^- > Br^- > Cl^- > py > RNH_2 = NH_3 > OH^- > H_2O$$

(7.2)

The *trans* effect or *trans* influence of a group coordinated to a metal (mainly d^8 SP-4) is the tendency of that group to direct an incoming ligand to the *trans* position. The outcome of the substitution reaction is thus the result of a competition between all ligands in such a complex. The terms *trans* effect and *trans* influence have been introduced in order to distinguish thermodynamic and kinetic factors influencing the stereo control in these substitution reactions [6]. There is an extended literature on the *trans* effect, treating experimental results and theoretical models. It is interesting that a simple empirical rule such as this so-called *trans* effect, which enables chemists to predict with a high degree of reliability the outcome of reactions that take often place in several steps (if several ligands are substituted), seems to withstand a simple theoretical explanation [7].

The application of the *trans*-effect rule is straightforward. Examples of its practical consequences are given in Figure 7.1.

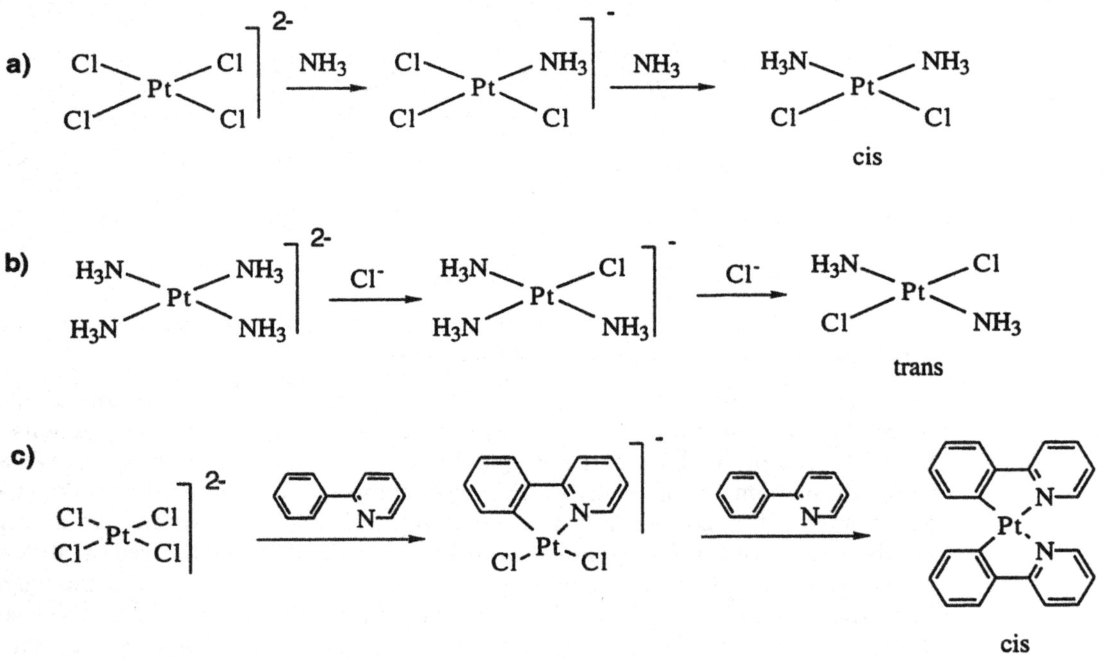

Figure 7.1
Three examples of applications of the *trans*-effect series. In each case, the stereospecificity for the synthesis of the diastereomer is near 100%

7.1.3 Rearrangements in Metal Complexes

The *trans* effect or *trans* influence discussed in above is probably the only *simple general rule* applicable to straightforward substitution reactions.[†] We therefore discuss other generalized acid–base reactions here under the title of 'rearrangements in metal complexes,' which has been borrowed from a review by Jackson and Sargeson [9] that also gives many mechanistic details of such reactions. A second review by Jackson [10] concentrates on inversions of configurations in ligand substitution processes. Rearrangement reactions in coordination compounds have been considered from various different points of view. Some are discussed in this chapter. Topological approaches will not be discussed here. The reader is referred to the literature [11–13].

Rearrangements can be divided, from a formal point of view, into several classes of transformations: *isomerizations* (I) do not change the stoichiometry (and the molecular weight, see p. 47) of a species under consideration, whereas *general rearrangements* (II) include substitution of ligands, reactions of ligands, etc. Isomerizations have recently been considered in a fairly detailed manner from a theoretical point of view, especially with respect to symmetry selection rules, by

---

[†]Much insight into the mechanism of substitution reactions, especially solvent exchange processes, has been gained in recent times through the application of NMR spectroscopy, especially the pressure dependence of NMR spectra [8].

Rodger and Schipper [14]. They distinguished two different classes of isomerizations: (I) isomerization between structures having the same point symmetry (called *homoconversions*), and (II) those of different point symmetry (called *heteroconversions*). We subdivide homoconversions further into two categories: *isoexchange reactions* (Ia) where the product is chemically identical with the reactant, and *racemizations* (Ib) where the product is the enantiomer of the reactant. Heteroconversions can also be called diastereomerizations. Racemization is a special case of *inversion*, which can also occur in general rearrangements, leading to non-isomeric products.

7.1.3.1   Isoexchange Reactions

A simple example of an isoexchange is, e.g., the Berry pseudo-rotation of TPY-5 complexes. Although it is a process where the 'product' and the 'reactant' are indistinguishable species, the transformation may be an observable process through the memory of the spin states of the nuclei, which are in non-equivalent positions. It is not easy to decide whether an isoexchange proceeds through one transition state only, or through an intermediate [14]. Moreover for reactions taking place in solution, it is difficult to determine what role the solvent plays in the conversion reaction. If solvent molecules participate appreciably in the transition state (or in an intermediate), the latter is not a polytopal isomer of the molecule under consideration but a new 'species' with different stoichiometry.

Molecules that undergo rapid isoexchange rearrangements are often called 'non-rigid structures' [15]. It is interesting that isoexchange seems to be mostly either a very rapid or a very slow process, depending on the coordination number. In general, rapid isoexchange occurs in five-coordinated species, and probably also in some complexes of coordination number >6, but it seems to be rare for T-4, SP-4, and OC-6 [16].

A somewhat different, but phenomenologically similar, isoexchange is possible in complexes with alterdentate ligands. An alterdentate ligand possesses, by definition [17], two or more *identical* coordinating sites. An interchange from one to another of these sites corresponds to an isoexchange process. Consider, for example, the two ligands ninhydrin anion and alloxane anion (Figure 7.2).

Both ligands of Figure 7.2 are radicals giving highly resolved EPR spectra [18,19]. The symmetry of the free ligands is C_{2v} in both cases, and the nuclei are pair-wise equivalent. Upon complexation at the symmetry equivalent sites A or A' (whereby the ligands acts as bidentate chelates), the symmetry is reduced to C_s. The mirror plane is coincident with the plane defined by the atoms of the alterdentate ligand. The nuclei in the ligand are now all non-equivalent, and the hyperfine coupling constants for the magnetic nuclei therefore different. Since the complex is dissolved in a coordinating solvent, the other coordination sites of the metal are occupied by solvent molecules. The isoexchange of a metal from site A to site A' (Figure 7.2) (which corresponds to a racemization in two-dimensional space) exchanges the nuclei in corresponding pairs, designated with primed and unprimed numbers, respectively. The memory of the spin states of the nuclei makes the exchange observable. Detailed EPR spectroscopic investigations (Von Zelewsky A., Moser E. M., and Wolny J., to be published) indicate that the exchange proceeds via an intermediate which also has C_s symmetry (Figure 7.3), but this time with the mirror plane perpendicular to the ligand molecule. The coordination geometry,

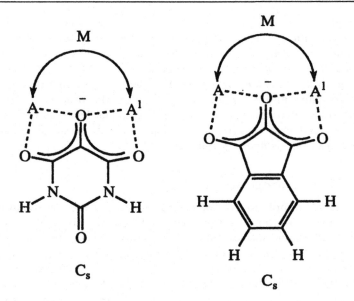

Figure 7.2
Isoexchange (and homoexchange) processes in alterdentate ligands. The species before and after the rearrangement are identical

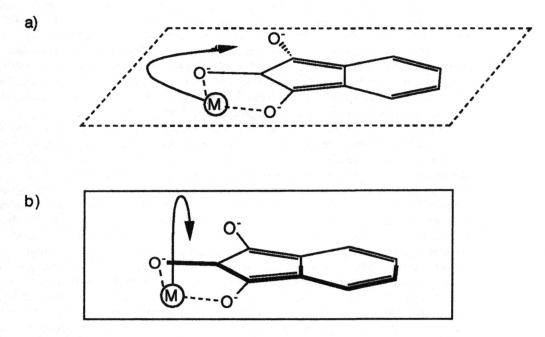

Figure 7.3
Two possible pathways for an isoexchange process

including the solvent molecule ligands, for the intermediate is not known, however. It is highly probable that in this case the solvent plays an important role in the transition state and/or the intermediate.

A series of complexes in which terpy acts as an alterdentate ligand have been reported recently for Pt^{IV}, Re^{I}, and Ru^{II} metal centers [20–23]. In all three cases fluxional motions occur, where the metal exchanges position between the two equivalent positions of terpy when it acts as a bidentate ligand. The most complete study was carried out with the complex $[PtI(CH_3)_3terpy]$. In this species, with a facial configuration of the three methyl ligands (Figure 7.4), three dynamic intramolecular processes can be observed by NMR spectroscopy, namely

- Pendant pyridine rotation, i.e. a purely conformational change.
- Exchange of the metal from coordination site A to the equivalent coordination site B, i.e. an isoexchange process.
- Scrambling of the methyl groups, which is a heteroexchange since the three CH_3 ligands are in non-equivalent positions.

7.1.3.2   Racemizations

Intramolecular racemizations.   An important class of isomerization reactions is racemizations, which can be distinguished from genuine inversion reactions, although racemization involves inversion. In racemization, the sum of the ee-values of the chemical compounds which participate in the reaction starts from some non-zero value and approaches zero during the reaction, whereas in inversion the ee-value changes sign if the chirality descriptors are chosen appropriately. Often, racemizations, inversions, and substitutions are closely related processes. We divide the discussion into these three subjects for the sake of facilitating the comprehension of the various phenomena. Substitution and inversion reactions will be discussed in the subsequent sections.

In general, racemization will take place if the enantiomer of a chiral compound can be reached through an activation barrier that is sufficiently low to make the process observable. Either there is just one transition state, or there are one or several intermediates along the reaction coordinate. Racemization is, from a thermodynamic point of view, always a spontaneous reaction. In a non-chiral environment, the enthalpy for racemization is $\Delta H^\circ = 0$, and the entropy corresponds to the purely statistical value of $\Delta S^\circ = R\ln2$, yielding a Gibbs free energy difference of $\Delta G^\circ = -1.72\,\mathrm{kJmol^{-1}}$ between the racemate and either of the two pure enantiomers. The time scale on which the process occurs for racemization can vary over a very large range. It can be extremely short, preventing the isolation of one enantiomer, e.g. in the case of amines $NRR'R''$, or very long, preventing observation of racemization of a pure enantiomer.

It is not necessary that the molecular system becomes achiral anywhere along the reaction coordinate. An OC-6 [Mabcdef] complex, for example, can change its configuration (i.e. go from C to A), through an intramolecular exchange of two ligands without ever having an achiral structure. As can be seen easily from the Bailar Tableau (Table 5.2), only an interchange of a *trans* pair leads to the enantiomer. An exchange of the ligands in a *cis* pair leads to a diastereomer (Figure 7.5).

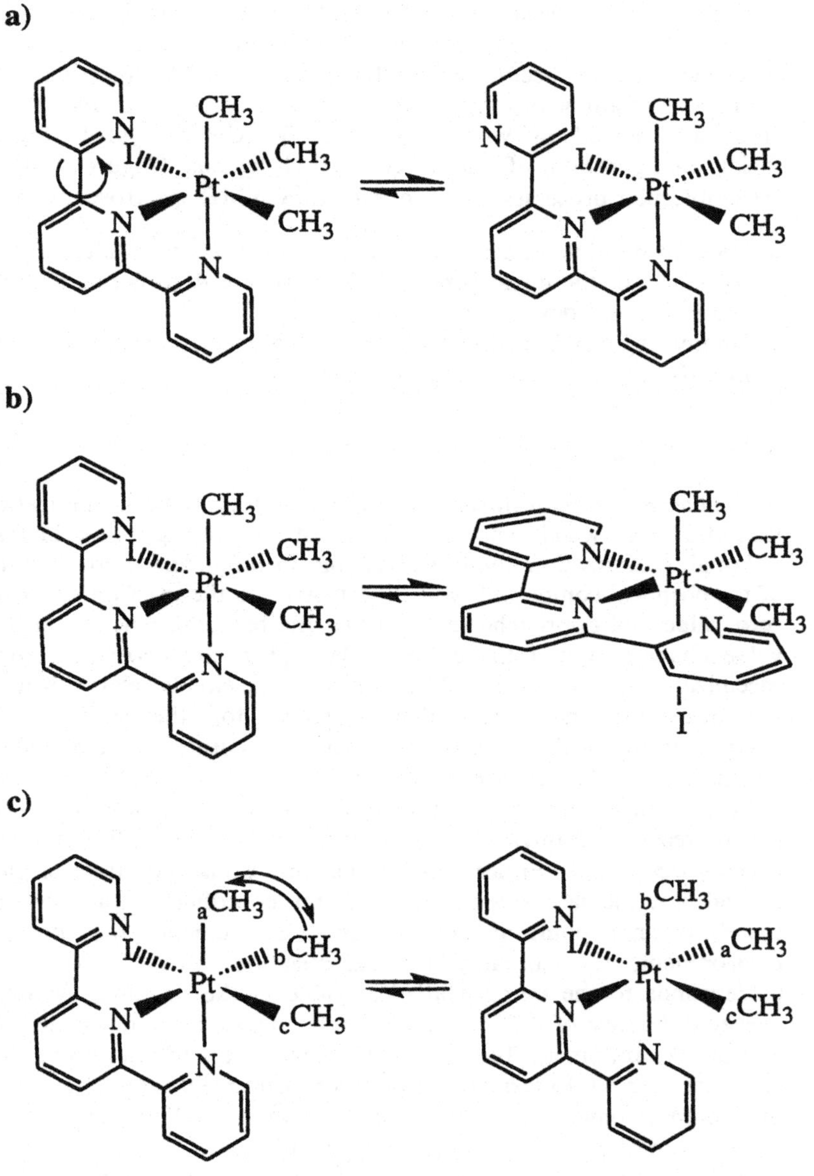

Figure 7.4
Three different rearrangement processes in the [PtI(CH$_3$)$_3$terpy] complex: (a) conformational change within the terpy ligand; (b) isoexchange (a homoexchange process) within the alterdentate terpy ligand; (c) scrambling of the non-equivalent methyl groups, a heteroexchange process

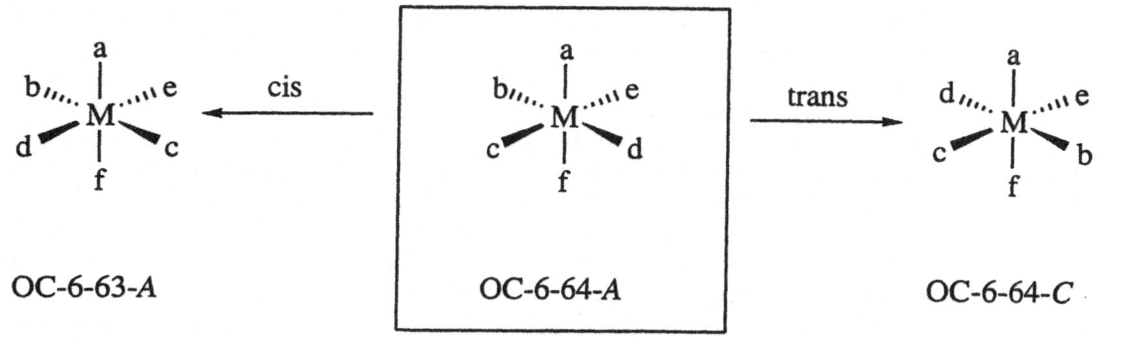

Figure 7.5
Interchange process in an OC-6 complex

Chiral molecules can be classified according to their behavior upon transformation of one enantiomer into the other [24]. Class I skeletons [Figure 7.6(a)] necessarily pass through an achiral intermediate, whereas for those of class (II) [Figure 7.6(b)] such a transition does not necessarily pass through an achiral state. Only T-4 and TB-5 belong to class I, whereas all the higher coordination polyhedra belong to class II) (Ref.13, p.45).

By far the largest number of chiral compounds of transition metals studied so far are those with chelate ligands, which can in most cases be unambiguously designated by the chirality descriptors Δ/Λ. The simplest case is the D_3-symmetric OC-6

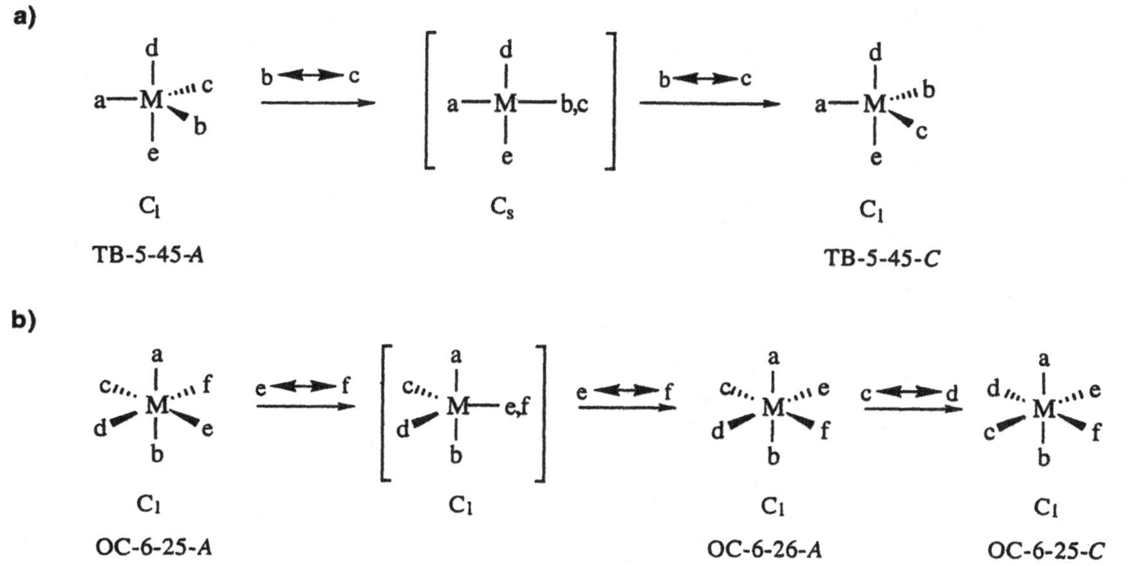

a)

b)

Figure 7.6
Examples of chiral species of classes I (TB-5) and II (OC-6). In TB-5 any continuous deformation that leads to the enantiomer goes through an achiral intermediate. In OC-6 deformations exist, where the object is always chiral along the reaction coordinate

complex $[M(A{}^{\wedge}A)_3]$ which can racemize in four different concerted intramolecular rearrangements, where the bidentate ligands remain coordinated in a symmetric manner to the central metal [25]. Two of them (*push through* and *cross over*, respectively) are very unlikely to occur for energetic reasons. The two others proceed both through a TP-6 transition state or intermediate (again, it is difficult to decide whether it corresponds to a local minimum or to a saddle point on the energy surface). The *Ray–Dutt* twist, also called *rhombic* twist [26] [Figure 7.7 (a)], corresponds to a transformation Δ-OC-6 $(D_3) \rightarrow$ TP-6 $(C_{2v}) \rightarrow \Lambda$-OC-6 (D_3). The *Bailar* twist, also called *trigonal* twist [27] [Figure 7.7 (b)], has a transition state/ intermediate of higher symmetry: Δ-OC-6 (D_3) TP-6 $(D_{3h}) \rightarrow \Lambda$-OC-6 (D_3).

In the Bailer twist, all three chelates remain equivalent during the racemization,

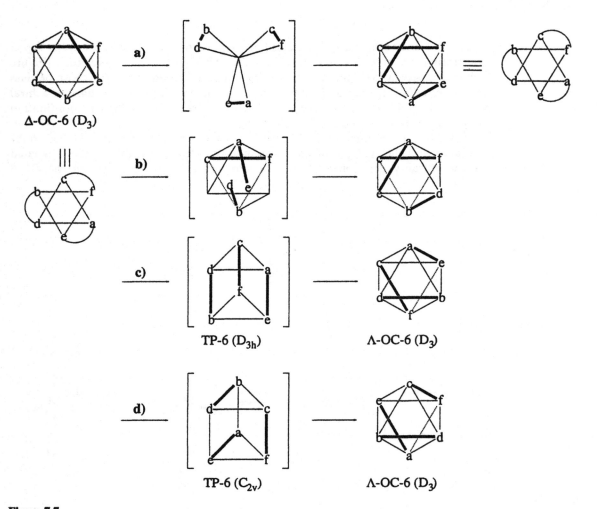

Figure 7.7

Four possibilities for intramolecular racemization for tris chelate OC-6 complexes: (a) push trough; (b) cross over; (c) Bailar twist; (d) Ray–Dutt twist

whereas in the Ray–Dutt twist the bidentate ligands become inequivalent $(2 + 1)$. For the latter, three energetically degenerate reaction paths must therefore occur for symmetry reasons (for tris-homoleptic OC-6 complexes with bidentate ligands). If an OC-6 complex racemizes intramolecularly, the mechanism will depend mainly on the chelate bite. A small bite will prefer a Bailar twist, a large bite a Ray–Dutt twist [25]. Actual studies in most cases, led to the conclusion that a Ray–Dutt twist, especially in solid-state reactions, is more likely to occur.

Intermolecular racemizations. Many racemization reactions of OC-6 complexes evidently proceed through intermolecular reaction mechanisms, some of them with and some without substitution of one of the ligands. An example of the latter is, e.g., the complex $[(NH_3)_4Co(sarcosinato)]^{2+}$, which is chiral solely by virtue of the stereogenic coordinated nitrogen ligand atom of the sarcosine anion. It does not racemize in slightly acidic solution. Increasing pH values lead to racemization of the complex, indicating clearly that an acid–base equilibrium becomes important for this reaction (Ref. 9, and references cited therein). The inversion of the configuration of nitrogen becomes rapid upon deprotonation of this ligand center, when it passes from T-4 to TPY-3 (Figure 7.8).

Acceleration of racemization through deprotonation has also been found in the

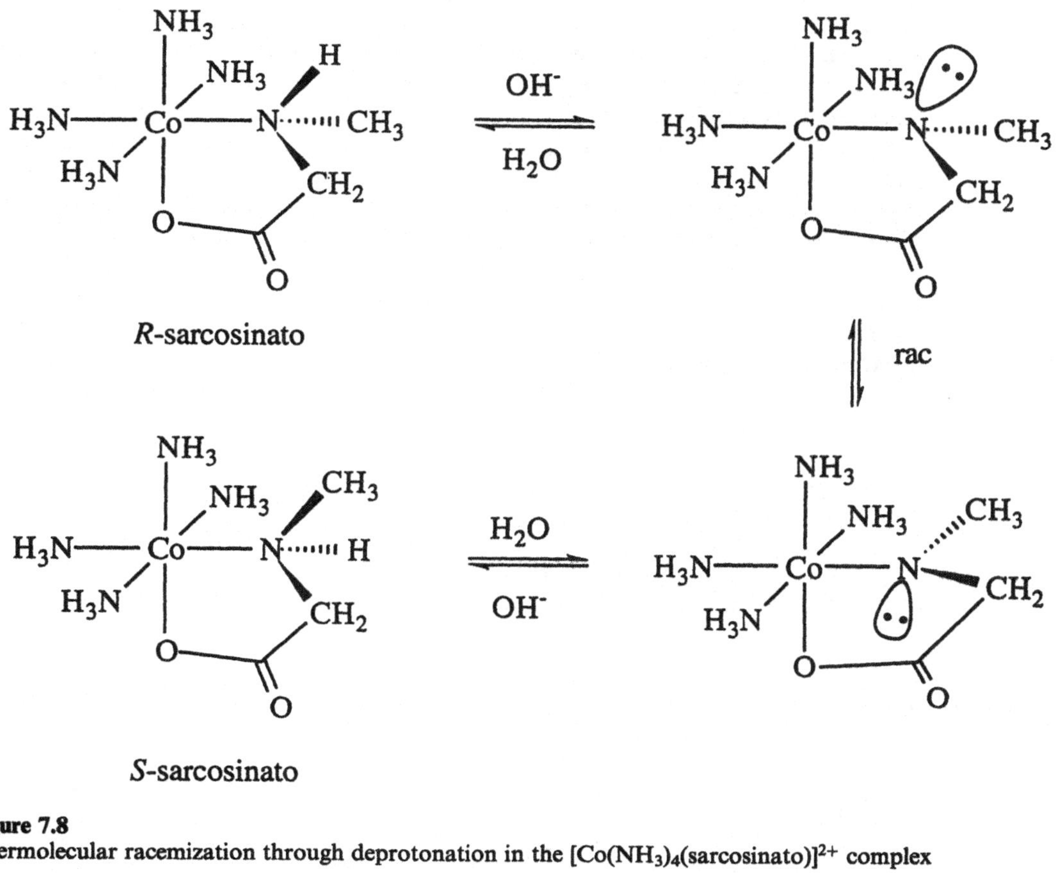

Figure 7.8
Intermolecular racemization through deprotonation in the $[Co(NH_3)_4(sarcosinato)]^{2+}$ complex

optically active complex $[Co(dien)_2]^{3+}$, configuration 32 [28]. Although a careful analysis was carried out, no complete certainty was achieved regarding the question of whether deprotonation and racemization are really synchronous processes. Similar racemizations with phosphine, arsine-, thio, and tellurio ligands have been investigated [9].

Detailed accounts of mechanisms of some isomerization processes, mainly in six-coordinate chelate complexes, have been given in two reviews [29,30], which consider also fast processes studied by NMR spectrosopy. The latter have also been treated thoroughly in several other publications [31,32].

7.1.3.3 Substitution and Rearrangement Reactions in OC-6

A large number of studies have been devoted to solvolysis, especially aquation reactions of the inert metal ions Cr^{III}, Rh^{III}, Ir^{III}, Pt^{IV}, Ru^{III}, Ru^{II}, and especially Co^{III}. For the last ion, an enormous amount of data on substitution reactions has been accumulated in the chemical literature. There are several reviews on the subject [30,33–43], and the reviews were reviewed by Sargeson and Jackson [9] in some depth. A large part of this work is focused on mechanistic problems and often the stereochemical aspects have not been considered in all the details. Our picture of the stereochemical course of substitution reactions in these OC-6 complexes is still being further developed, and it seems impossible at this moment to give a description which boils down to a few basic rules. Some generalizations can be formulated, but they should be taken *cum grano salis*: (i) for the inert metal centers mentioned above, most complexes will rearrange slowly relative to the rate of aquation, with the exception of Co^{III}; (ii) Co^{III} complexes often rearrange during aquation; and (iii) in many cases, the steric control of an aquation reaction depends strongly on experimental conditions, such as temperature. For example, the aquation of *cis*-$[Co(en)_2Br_2]^+$ gives temperature-dependent yields of *cis* and *trans* products [44].

Werner carried out a number of substitution and ligand transformation reactions on optically active cobalt complexes (Ref. 45, p. 368). He observed in several cases an inversion of the sense of rotation of the polarized light (at a certain wavelength). The reactions described are shown in (7.3).

a) $(-)[Co(en)_2Cl_2]^+ \xrightarrow{K_2CO_3} (-)[Co(en)_2CO_3]^+$

b) $(-)[Co(en)_2ClSCN]^+ \xrightarrow{NaNO_2} (-)[Co(en)_2NO_2SCN]^+$

c) $(-)[Co(en)_2NH_2O_2Co(en)_2]^{4+} \xrightarrow{reduction} (+)[Co(en)_2NH_2OHCo(en)_2]^{4+}$

d) $(-)[Co(en)_2NH_2O_2Co(en)_2]^{4+} \xrightarrow{HNO_2(aq)} (+)[Co(en)_2NH_2NO_2Co(en)_2]^{4+}$

$$(7.3)$$

Assuming, presumably correctly, that the absolute configuration could not change during these reactions, Werner concluded that the sign of the optical rotation at one wavelength is not directly related to the absolute configuration of the complex. What Werner's results also show is the fact that many substitution reactions on cobalt complexes proceed with retention of configuration. Jackson and Sargeson [9] have reviewed the stereochemical course of the stepwise substitution reaction of the two chloride ligands in $[Co(en)_2Cl_2]^+$.

Experiments carried out under a variety of different conditions show that these substitutions often yield a mixture of isomers, indicating partial retention, partial racemization, and some rearrangement (cis–trans). Predominant, however, is retention of configuration [46,47]. The acid-catalyzed aquation of $[Co(en)_2(a)(NO_2)]^{n+}$ to $[Co(en)_2(a)(H_2O)]^{(n+1)+}$ occurs with retention of configuration [48]. In a general way, it can be stated that under certain conditions retention of configuration, with little rearrangement, is the predominant course of substitution reactions of cobalt(III) complexes. In some cases this is also true for substitution reactions on other metals. It was recently shown that some ruthenium(II) complexes undergo substitution reactions with retention of configuration if two pyridine ligands are replaced by any bidentate chelate. This has been exploited for the synthesis of isomerically pure polynuclear complexes [49,50].

An interesting enantioselective substitution reaction yielding chiral coordination species in non-racemic composition is provided by two reactions (7.4) described in the early literature [51].

$$\Delta,\Lambda\text{-}[Co(en)_2(d\text{-}(-)\text{-tartrate})]^+ + en \longrightarrow \Lambda\text{-}[Co(en)_3]^{2+} + [d\text{-}(-)\text{-tartrate}]^-$$

$$\Delta,\Lambda\text{-}[Co(en)_2(d\text{-}(-)\text{-tartrate})]^+ + 2\,NO_3^- \longrightarrow \Lambda\text{-}[Co(en)_2(NO_3)_2] + [d\text{-}(-)\text{-tartrate}]^-$$

$$(7.4)$$

A stereoselective substitution reaction of an enantiomerically pure metal complex with a racemic ligand was used to separate the enantiomers of a chiral ligand [52] (7.5). In the same publication, it is shown that a racemic metal complex can be partially resolved by stereospecific reaction with an enantiomerically pure ligand (7.6).

$$\Lambda\text{-}[Co(edta)]^- + rac\text{-pn} \longrightarrow \Lambda\text{-}[Co(S\text{-}(-)\text{-pn})_3]^{3+} + R\text{-}(+)\text{-pn}$$

$$(7.5)$$

Bailar and co-workers [53] showed that $[Ru^{II}(bpy)_3]^{2+}$ and $[Os^{II}(bpy)_3]^{2+}$, which were first resolved into their enantiomers by Burstall et al. [54,55], can be enantioselectively synthesized from ruthenium and osmium complexes in higher oxidation states and bpy in the presence of tartrate or sucrose, yielding relatively modest ee-values. It is assumed that the chiral auxiliaries serve both to direct the stereochemistry of the reaction and to reduce the central metal.

$$\textit{rac-}[\text{Co(edta)}]^- + 3\,R\text{-(+)-pn} \longrightarrow \Delta\text{-}[\text{Co}(R\text{-(+)-pn})_3]^{3+} + \Lambda\text{-}[\text{Co(edta)}]^-$$

$$(7.6)$$

Inversion at OC-6 centers. Before describing experimental results of reactions proceeding with inversion, a short theoretical discussion is appropriate. Inversion as a single-step transformation of one enantiomer into its mirror image form is thermodynamically forbidden (at least in a non-chiral environment). Only racemization is allowed. Inversion, however, is not restricted to isomerization. The term inversion refers always to a single chirality center in a molecule. Such a chirality center can be characterized by either one of the descriptor pairs R/S, C/A, and Δ/Λ (or λ/δ). If a chirality center is maintained during a chemical reaction, it can have either the same or the opposite chirality designation in the reactant, and in the product, respectively. Change in the chirality descriptor *does not necessarily* signify inversion. This is only the case for the Δ/Λ or $\overrightarrow{\Delta}/\overrightarrow{\Lambda}$ pair, but neither for R/S nor for C/A pairs. For the latter a suitable presentation has to be chosen in order to find out whether a reaction proceeds under inversion or retention of configuration. Figure 7.9 gives examples of the various possibilities.

Inversions at T-4 centers belong to the field of classical organic substitution reactions. This class of reactions is normally referred to as the 'Walden inversion.' Bailar and Auten [56] discovered in 1934 an inversion reaction of an octahedral complex. Such reactions are called Bailar inversions and they were reviewed thoroughly by Jackson [10] in 1986. There are not many reactions which have proven to be genuine Bailar inversions, but the reaction originally investigated by Bailar and Auten undoubtedly proceeds by inversion in strongly basic homogeneous solution (7.7).

The immediate mechanistic problem connected with this reaction is the question of whether it is a two-step or a one-step process. It can be carried out under conditions so that an 'intermediate' $[\text{Co(en)}_2\text{Cl(OH)}]^+$ can be observed. Although it was originally thought that the inversion takes place in the first step [57], detailed investigations showed later [10] that the *stepwise* substitution of Cl^- by OH^- proceeds by *partial racemization* and slight rearrangement, but *not* by *inversion* (7.8).

The inversion process must therefore be a *one-step* reaction. There have been speculations about the intermediate/transition state of this reactions, and Jackson proposes a species with CN 4.[†] Bailar and co-workers [61–64] also observed inversion under certain conditions for Co^{III} complexes with multidentate ligands, such as trien.

Ambidentate ligands can rearrange from one to another linkage isomer if the thermodynamically less stable form can be synthesized. This is the case, e.g., for

---

[†]A straightforward explanation of the experimental findings is the assumption of a substitution of the two Cl^- ligands by an OH^- dimer, $(\text{O}_2\text{H}_2)^{2-}$, in a *bimolecular* step through a transition state of increased coordination number. Inversion is much more likely to occur via a transition state with *increased* CN, as is shown in T-4 by the well known Walden inversion. Anion dimers have been found in aqueous solutions of Cl^- [58–60]. Owing to its hydrogen-bonding capability, OH^- is even more likely to form dimers than Cl^-. Numerical estimations of the concentration of such dimers show that this possibility should be taken seriously into consideration.

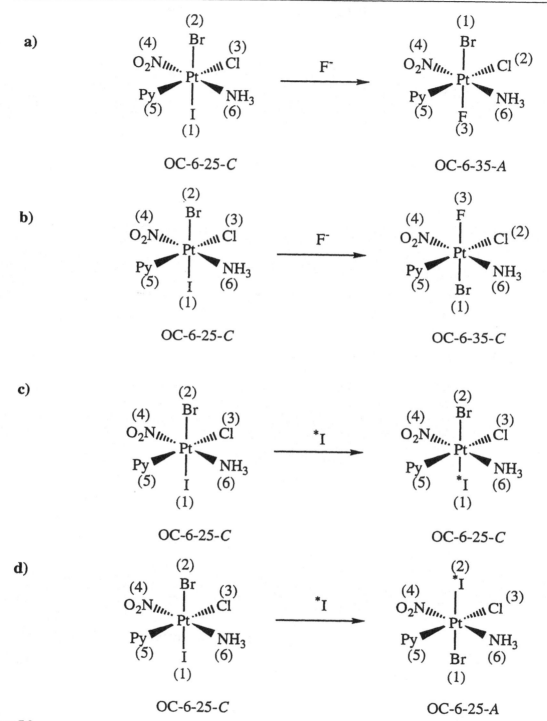

Figure 7.9
Examples of true inversions and pseudo-inversions

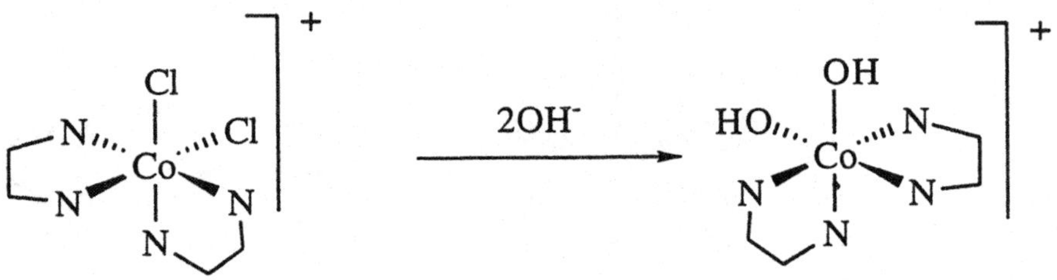

$$\Delta-cis \qquad\qquad \Lambda\text{-}cis$$

$$(7.7)$$

$$\Delta\text{-}cis[Co(en)_2Cl_2]^+ \xrightarrow{OH^-} [Co(en)_2ClOH]^+ \xrightarrow{OH^-} \Delta,\Lambda\text{-}cis[Co(en)_2OH_2]^+ + trans[Co(en)_2OH_2]^+$$

$$(7.8)$$

$[M(NH_3)_5(\underline{O}NO)]$, with $M = Co^{III}$, Rh^{III}, Ir^{III}, and Pt^{IV}, which all isomerize to the thermodynamically more stable nitro form $[M(NH_3)_5(\underline{N}O_2)]$ [7].

7.2   Addition and Elimination Reactions

Addition and elimination reactions in coordination chemistry can be defined as processes where the coordination number of a given metal center increases or decreases, respectively. The best known examples are oxidative additions (OA) and reductive eliminations (RE) (7.9).

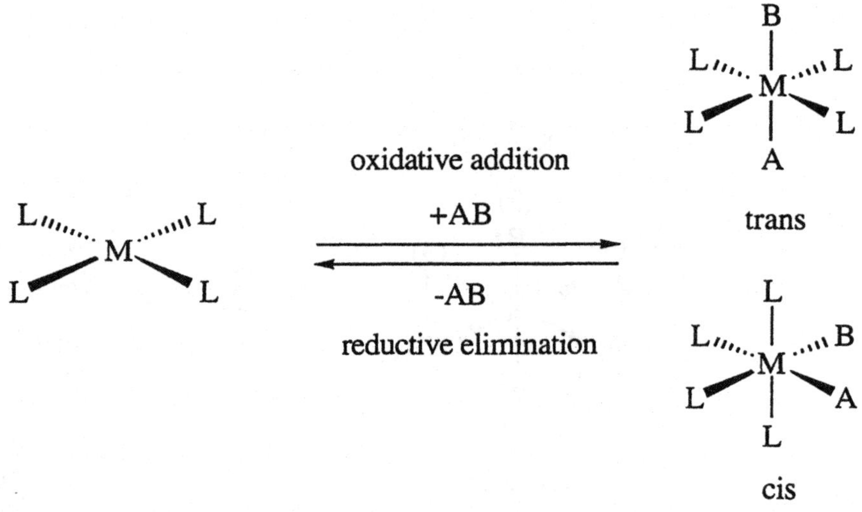

$$(7.9)$$

In most cases the two 'ligands' L are bound covalently, forming a molecule L_2 (which can be either A_2 or AB), before the addition takes place, or after the elimination. In the metal complexes they become anionic ligands. Therefore, the oxidation number of the metal increases by two units in the addition reaction and it decreases by two units in the elimination reaction. These reactions are therefore called *oxidative addition* and *reductive elimination*, respectively. In most cases, the SP-4 complexes have d^8 metal centers (Pd^{II}, Pt^{II}, Rh^I, Ir^I), yielding the corresponding d^6 OC-6 species with the higher oxidation numbers.

The stereochemistry of such reactions has been studied for many complexes with various metals and different types of ligands. Ir^I complexes with three different monodentate ligands, the so-called Vaska complexes, have been thoroughly studied.

A complex of the Vaska type $[Ma_2bc]$ yields, as an addition product, $[Ma_2bcde]$ with a molecule (de). The latter can be, e.g., H_2, Cl_2, or RX. The addition product can occur, in principle, in 15 stereoisomeric forms of which three are non-chiral and six exist as enantiomeric pairs. In reality OA is generally fairly stereospecific. Only a small number of isomers are formed. Two modes of addition have been observed: *cis*-addition with non-polar molecules such as H_2, and *trans*-addition with alkyl halides. In both modes, minimum rearrangement of the ligands originally present in the SP-4 d^8 complex is observed.

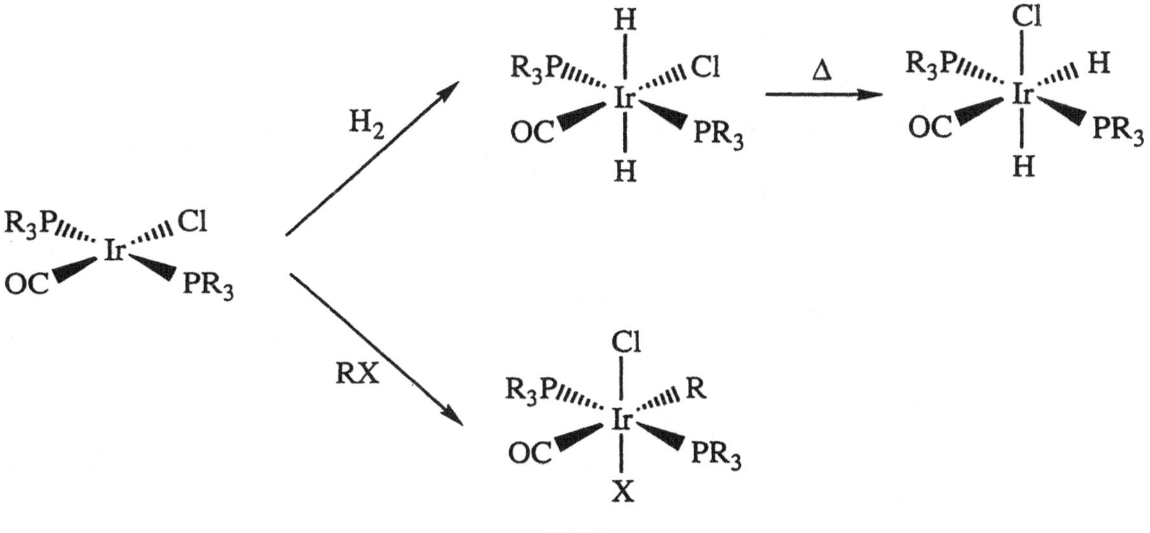

(7.10)

The *trans* product, obtained in the reaction with RX (7.10), can thermally isomerize to the *cis* product, showing that the addition reaction yields in this case the thermodynamically less stable compound. A detailed mechanistic study of an OA reaction has been carried out [65,66] with $[Pd(CO)(PPh_3)_3]$ as the SP-4 d^8 complex, and (S)-$[CHBr(CH_3)(Ph)]$ as the added molecule. This reaction leads through a sequence of OA, RE, and alkyl migration to a complex

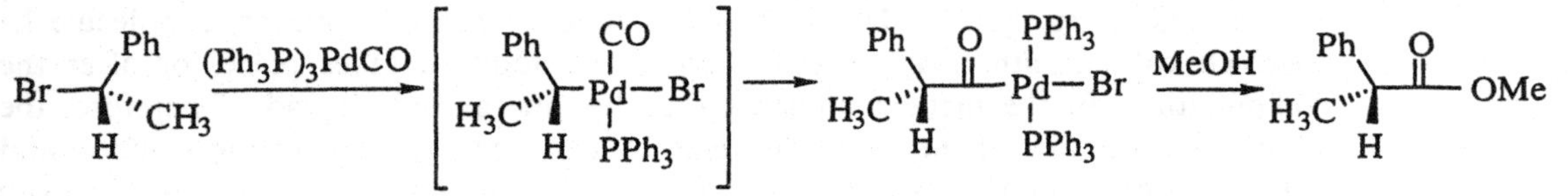

Figure 7.10
Sequence of OA, RE, and alkyl migration in a Pd complex

[PdBr(PPh$_3$)$_2$(COCH(CH$_3$)(Ph)], where the absolute configuration of the chiral carbon center has been genuinely inverted (Figure 7.10).

OA of Cl$_2$ to [Pt(en)$_2$]$^{2+}$ seems to yield exclusively the complex *trans*-[Pt(en)$_2$Cl$_2$]$^{2+}$ [67]. *cis*-[Pt(en)$_2$Cl$_2$]$^{2+}$ has to be prepared by another sequence of reactions [68].

OA reactions of SP-4 complexes with cyclometalated ligands have been studied in detail recently [69–71]. It was shown that some oxidative reactions with cyclometalating ligands occur only photochemically (POA), whereas others are sufficiently rapid to be observed also thermally (TOA) (Figure 7.11). Again, out of the numerous isomers possible, only a small number of them are formed during these reactions.

7.3   Formation of Supramolecular Species

Supramolecular chemistry, the chemistry beyond the molecule and the covalent bond [72], is a subject where stereochemical considerations are fundamental for understanding natural systems and for creating artificial structures. Chemistry of life is developing more and more towards an understanding of supramolecular phenomena and the construction of molecular devices will certainly also require the application of concepts developed in supramolecular chemistry [73]. Coordination compounds, especially those of transition metal ions, are of great interest in supramolecular chemistry because they can perform a large number of functions. In this section, a brief overview of some aspects of this rapidly developing field is given.

An aspect of supramolecular chemistry which has been under investigation for some time is the interaction of those species in a solid that are more or less independent in solution. Two different kinds of intermolecular interactions are particularly important in this respect:

- the interactions between the two enantiomeric molecular forms of a chiral compound in a solid;
- the diastereomeric interactions of different enantiomers in general.

The first of these two points is important for the question of whether a racemate crystallizes as a *racemic compound* or as a *racemic mixture* (conglomerate crystallization). In some series of compounds an understanding seems to emerge of why some compounds form mixtures whereas others, although seemingly closely related, form compounds. This is the case, for example, for

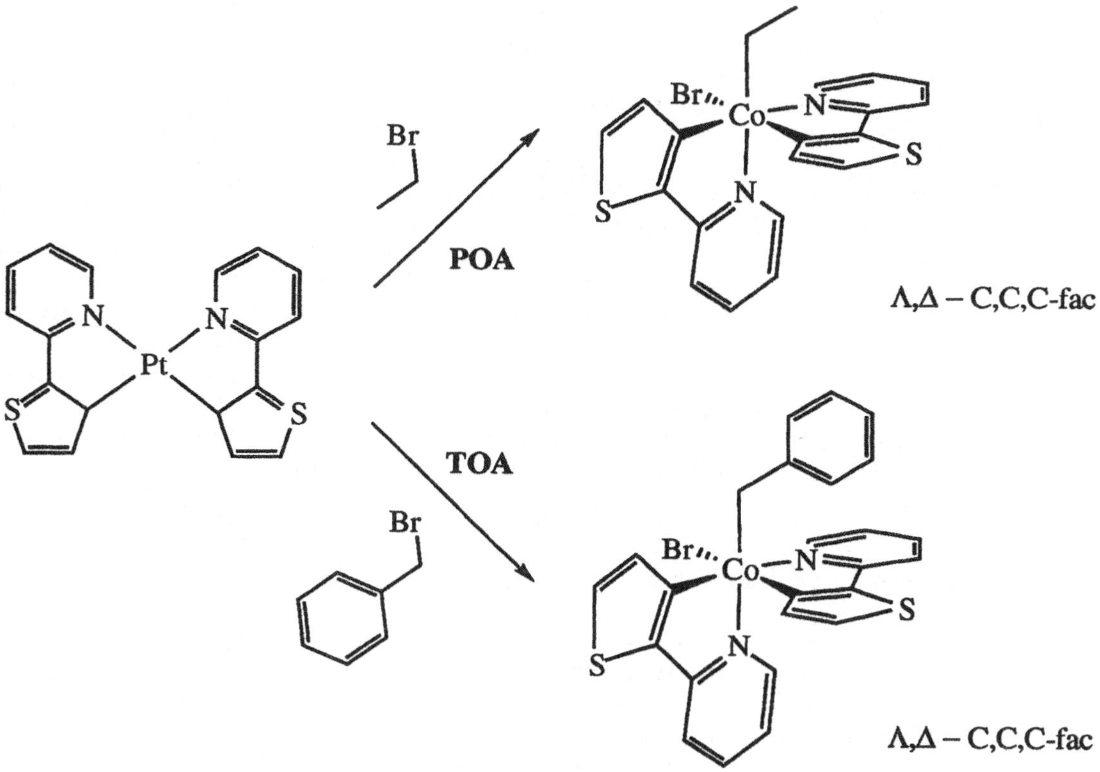

Figure 7.11
Examples of thermal oxidative addition (TOA) and photochemically induced oxidative addition (POA) in a cyclometalated SP-4 Pt(II)/OC-6 Pt(IV) complex. The reactions proceed in a highly stereospecific manner [71]

$(H_3O^+)[Co(en)_2(ox)]Cl_2 \cdot H_2O$, which forms centrosymmetric crystals containing both enantiomers as infinitely hydrogen-bonded helices of homochiral $[Co(en)_2(ox)]^+$ cations [74]. Adjacent helices of opposite chirality are held together by hydrogen bonds between the water molecules and the chloride ions and the NH_2 protons of the cations. The same cation in a compound of stoichiometric composition $[Co(en)_2(ox)]Cl \cdot 4H_2O$ forms analogous helices with the difference that they are all of the same chirality in one crystal, thus forming a racemic mixture. Whether it will become possible to predict reliably the occurrence of the phenomenon of spontaneous separation of a racemate upon crystallization (known since Pasteur) in general will depend whether the subtle differences of the intermolecular interactions can be understood in a sufficiently precise way. It seems that the realization of this aim is still in the rather distant future.

Similar considerations apply to the problem of diastereomeric interactions between an enantiomer and a pair of enantiomers in a racemate. Intermolecular diastereomeric interactions have been the basis for most resolutions of racemates carried out over decades. It seems that often intuition, or perhaps just diligence or serendipity, have led chemists to chose the ideal enantiomer for a given racemate to be resolved. It is hardly conceivable that Werner chose the anion of camphorsulfonic acid as a resolving agent for the Co–hexol complex on rational grounds. Eighty years later it was shown by Bernal[†] that the reason for the almost perfect resolvability is a sterically highly specific interaction between the homochiral OH bridges (p. 181) through hydrogen bridges with the SO_3 group of the chiral anion. It is difficult, however, to understand why $[Ru(1,10\text{-phen})_2(py)_2]^{2+}$ can easily be resolved by arsenyl tartrate, whereas the very similar $[Ru(2,2'\text{-bpy})_2(py)_2]^{2+}$ resists all attempts to be separated by the same resolving agent. Benzoyl tartrate, however behaves in just the opposite way [50]. Yoneda and Miyoshi [75] studied the structure of a number of diastereomeric crystalline compounds with some interesting results concerning their packing structures.

Highly interesting studies have been carried out in the field of supramolecular interactions between coordination compounds and DNA [76,77] These developments show the possibility of future applications of coordination chemistry in biology where stereochemical criteria are highly pertinent.

7.4   Reactions of Coordinated Ligands

In the genuine supramolecular species briefly discussed above, *weak interactions* between the coordination unit and other molecular entities are the important features. *Formation or breaking of covalent bonds* within the ligands is another important phenomenon in coordination chemistry. Here only a very brief outline is given of this extremely important field, where reactions take place at the periphery of the coordination units, i.e. at the coordinated ligands themselves.

The simplest reaction taking place at a coordinated ligand is protonation/ deprotonation. It can lead to stereochemical consequences, as in the case of $[Co(NH_3)_4(sarcos)]^{2+}$, discussed earlier, where racemization at the coordinated stereogenic nitrogen occurs upon deprotonation. In the case of chelated amino acids, deprotonation of the stereogenic carbon center leads to mutarotation [78].

Protonation/deprotonation is an important reaction, influencing the stereochemical course of many processes in organometallic chemistry. The discussion of such reactions is outside the scope of this book.

Dehydrogenation/hydrogenation has been investigated from the point of view of stereochemical changes. The oxidation of coordinated 2-(1-aminoethyl)pyridine (Meampy) in $[Ru(bpy)_2(Meampy)]^{2+}$, which forms the two chiral diastereomers $\Lambda S/\Delta R$ and $\Delta S/\Lambda R$, shows that (a) one of the diastereomers ($\Lambda S/\Delta R$) is more rapidly oxidized and (b) that the configuration of the central metal is retained upon oxidation [79]. The latter findings were confirmed (Jandrasics, E., Wolny, J. and Von Zelewsky, A., in preparation) for the oxidation of $\Delta\text{-}[Ru(bpy)_2(S,S\text{-}DACH)]^{2+}$ to the diimine (Figure 7.12).

The back-reaction, performed with a hydridic agent ($[BH_4]^-$), retains probably

---

[†]Communicated by I. Bernal at the Coordination Chemistry Centenial Symposium (C3S), 205th National Meeting of the American Chemical Society, Denver, CO, 1993.

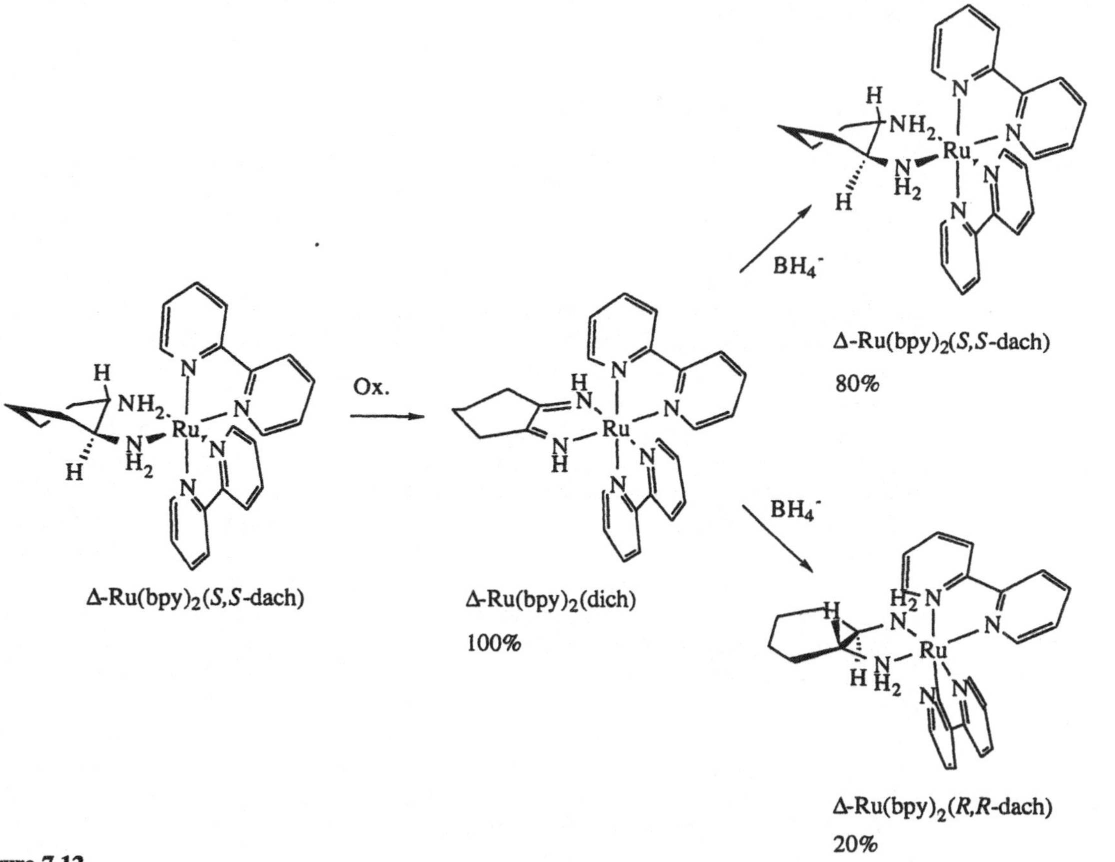

Figure 7.12
Sequence of hydrogenation/dehydrogenation of a coordinated diamine ligand (dach). The dehydrogenation takes place with full retention of configuration; the hydrogenation is partially diastereoselective

completely the configuration at the metal center and it is partially stereoselective with respect to the formation of the chiral centers at the ligand, yielding Δ-$[Ru(bpy)_2(S,S\text{-}DACH)]^{2+}$ and Δ-$[Ru(bpy)_2(R,R\text{-}DACH)]^{2+}$ with a de (diastereomeric excess) of 60% for the Δ-S,S diastereomer.

A large class of transformations are the reactions employed in the so-called template syntheses, discussed earlier in this book [80,81]. The formation of many macrocyclic ligands, of cage compounds, and of the catenands and knotted compounds is due to the template effects. It is predictable that this type of chemistry will be very important in the future for the construction of highly complex molecular assemblies.

Finally, the stereochemistry of reactions of coordinated ligands is of utmost importance in bioinorganic chemistry. Many enzyme reactions take place at metallic coordination centers. Generally such reactions are highly stereospecific. The knowledge of these processes has increased enormously during the past two decades, yet many details remain to be discovered.

7.5   Electron Transfer Reactions

Electron transfer reactions between coordinated species in solution have been studied intensely for the past four decades. For a reaction of the type

$$[A^{red}] + [B^{ox}] \rightarrow [A^{ox}] + [B^{red}]$$

where $[A^{red}]$, $[B^{ox}]$, $[A^{ox}]$, and $[B^{red}]$ stand for metal complexes forming two red/ox pairs (most studies deal with one-electron transfer reactions), two basic types of reaction mechanisms have been established [82]. In *outer-sphere reactions*, the coordination spheres of both partners remain intact during the electron transfer process, whereas in *inner-sphere reactions* a ligand bridge is established between the metal centers. Inner-sphere reactions therefore always involve also some rearrangement of the coordination sphere in at least one of the participating complexes. The bridging ligand may, or may not, be exchanged in the reaction. Energetics of electron transfer reactions have been studied intensively from the theoretical point of view in the past decades [83,84].

The stereochemical aspects of electron transfer reactions have more recently been explored also. Of particular interest is the question of how stereoselective electron transfer reactions are. The present knowledge in this field has recently been reviewed [85–87]. The interested reader is referred to these publications, where the mechanistic aspects are discussed in some detail.

Most observations on the stereoselectivity of electron transfer reactions have been made with outer-sphere reactions of the type in (7.11).

One of the reactants is used in the enantiomerically pure or enriched form, the other as the racemate. Product analysis, e.g. by circular dichroism spectroscopy, of the complex originally present as the racemate shows whether the reaction is stereospecific (ee $\neq$ 0) or non-stereospecific (ee = 0). The reaction will be stereospecific if diastereomeric encounter complexes of the type $\{\Lambda\text{-}[OC\text{-}6, ML]^{red/ox}/\Lambda\text{-}[OC\text{-}6, ML']^{red/ox}\}$ and $\{\Lambda\text{-}[OC\text{-}6, ML]^{red/ox}/\Lambda\text{-}[OC\text{-}6, ML']^{red/ox}\}$ differ sufficiently in energy to produce an observable effect. It is very important to choose the complexes judiciously, since several criteria have to be met by the participating complexes. Notably, stereoselectivity can become unobservable if the product racemizes rapidly by a self-exchange electron transfer process [86].

The first examples where stereoselectivity was established without doubt [88] are shown in (7.12).

Co^{2+} and en form a racemic equilibrium mixture, owing to the high lability of the Co^{2+} coordination center. There have been many other reports on similar outer-sphere electron transfer reactions [85], indicating that subtle differences in intermolecular interactions (ion pairing, hydrogen bonding, and stacking interactions) are decisive for the sign and the magnitude of the stereoselectivity in outer-sphere electron transfer reactions.

Inner-sphere electron transfer reactions have, in certain cases, been shown to be stereoselective too. An elegant investigation, involving chiral OC-6 complexes, makes use of the pentadentate ligands mentioned in Chapter 5, Section 5.5 [85] (p. 142). As an example, the reduction of $[Co(bamp)(H_2O)]^+$ by various iron(II) complexes with similar ligands shows stereoselectivity in some cases [89, 90]. This reaction type is pertinent as a model for electron transfer reactions in biochemical systems.

$$\Lambda\text{-}[OC\text{-}6,\ ML]^{red/ox} + \Delta,\Lambda\text{-}[OC\text{-}6,\ ML']^{ox/red} \longrightarrow \Lambda\text{-}[OC\text{-}6,\ ML]^{ox/red} + \Delta,\Lambda\text{-}[OC\text{-}6,\ ML']^{red/ox}$$

stereospecific

$$\Lambda\text{-}[OC\text{-}6,\ ML]^{red/ox} + \Delta,\Lambda\text{-}[OC\text{-}6,\ ML']^{ox/red} \longrightarrow \Delta,\Lambda\text{-}[OC\text{-}6,\ ML]^{red/ox} + \Delta,\Lambda\text{-}[OC\text{-}6,\ ML']^{ox/red}$$

non stereospecific

(7.11)

a)

$$\Delta\text{-}[Os(bpy)_3]^{3+} + \Delta,\Lambda\text{-}[Co(EDTA)]^{2-} \longrightarrow \Delta\text{-}[Os(bpy)_3]^{2+} + \underset{52.5\ \%}{\Delta\text{-}[Co(EDTA)]^{-}} + \underset{47.5\%}{\Lambda\text{-}[Co(EDTA)]^{-}}$$

$S_2O_8^{2-}$

SO_4^{2-}

b)

$$\Delta\text{-}[CoEDTA]^{-} + \Delta,\Lambda\text{-}[Co(en)_3]^{2+} \longrightarrow \Delta\text{-}[CoEDTA]^{2-} + \underset{73\%}{\Delta\text{-}[Co(en)_3]^{3+}} + \underset{27\%}{\Lambda\text{-}[Co(en)_3]^{3+}}$$

(7.12)

7.6   Enantioselective Catalysis by Metal Complexes

A new book on the stereochemistry of the coordination chemistry of coordination compounds would be grossly incomplete without some notes about enantioselective catalysis, because the essential steps in these reactions most often take place within the coordination sphere of metal complexes. On the other hand, many aspects of this fascinating field, especially the mechanistic considerations, are outside the scope of the present treatment of stereochemistry.

Enantioselective catalysis, often also called asymmetric catalysis,[†] has been one of the fields in chemistry arousing very great interest during the past 30 years. Before ca 1966, enantioselective catalysis was considered to belong to the realm of living nature, where enzymes have produced chiral molecules in high enantiomeric purity since billions of years.

The rapid progress made subsequently in this field has been stimulated by the development of homogeneous catalysts, especially through metal complexes in general, and by a strong demand for enantiomerically pure compounds by the chemical industry.

A large percentage of the pharmaceuticals, estimated to represent a value of $18 billion ($1.8 \times 10^{10}$) per year [91] and an increasing share of agrochemicals, contain chiral molecules as active ingredients. Since all biochemistry proceeds in chiral systems, it is mandatory to use enantiomerically pure compounds as much as possible. The strong interest in this field is mirrored by a number of important recent publications about enantioselective catalysis [92–94]. The interested reader is referred to these publications for more detailed information, especially for applications. Here solely the principles will be discussed briefly.

Catalysis is, by definition [95], a purely kinetic phenomenon. Enantioselectivity can therefore only be achieved through an energetically preferred reaction pathway for one of the enantiomers from the chiral precursor to the chiral product. Reactions catalyzed by metal complexes are often multi-step processes following complicated reaction mechanisms. At least one of the transition states from the substrate to the chiral product must be sufficiently different in energy for the two enantiomers in order to achieve enantioselectivity, and we consider here only this step. There are, of course, many other conditions that must be fulfilled. For example, no other reaction step may lead to racemization of an intermediate obtained in enantiomerically enriched form (Figure 7.13).

Figure 7.13 gives schematically the reaction profile for the enantioselective step in a catalytic reaction. It is not only the difference in activation energy, $\Delta\Delta G^{\ddagger}$, which determines the enantioselective efficiency of the catalytic process. The reaction conditions strongly influence the outcome of the reaction. It is evident, for example, that the ee obtained is a function of several parameters, including time. In a closed system, ee(t) must approach 0 for $t \to \infty$, i.e. no enantioselectivity can be obtained if sufficient time is allowed to reach equilibrium. Therefore, a careful optimization of the reaction conditions is necessary in any practical case. Most

----

[†]*Asymmetric catalysis* is one of the numerous and often used misnomers in stereochemistry. Although chemists understand clearly what is meant with the term asymmetric catalysis, the more exact and self-explanatory *enantioselective catalysis* is to be preferred.

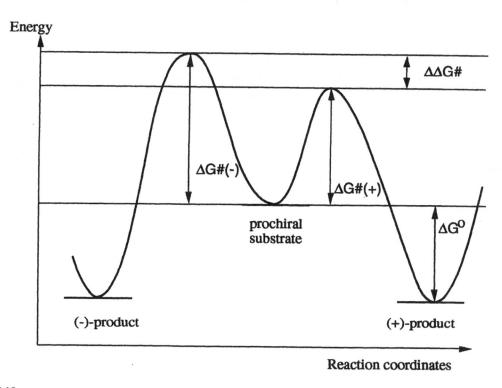

Figure 7.13
Simplified energy profile for an enantioselectively catalyzed reaction. The designation of the enantiomers (+) and (−) is arbitrary

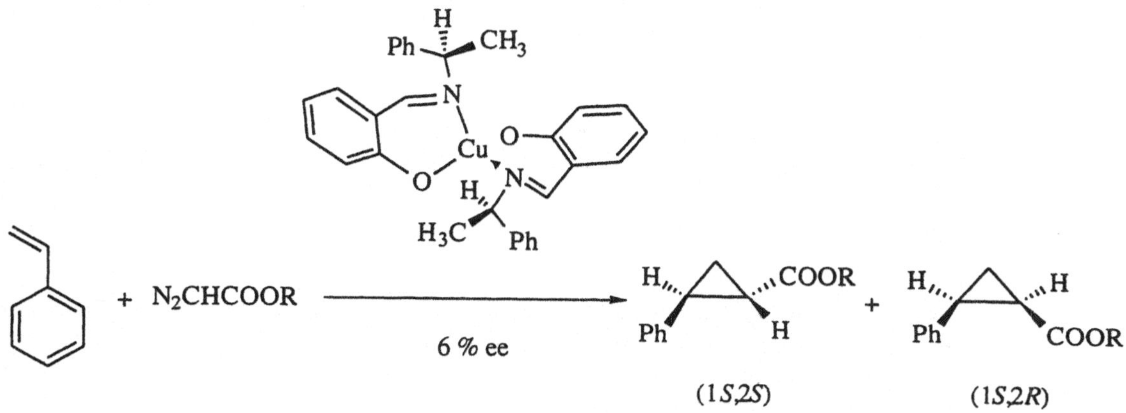

Figure 7.14
The first enantioselective catalytic reaction reported in the literature

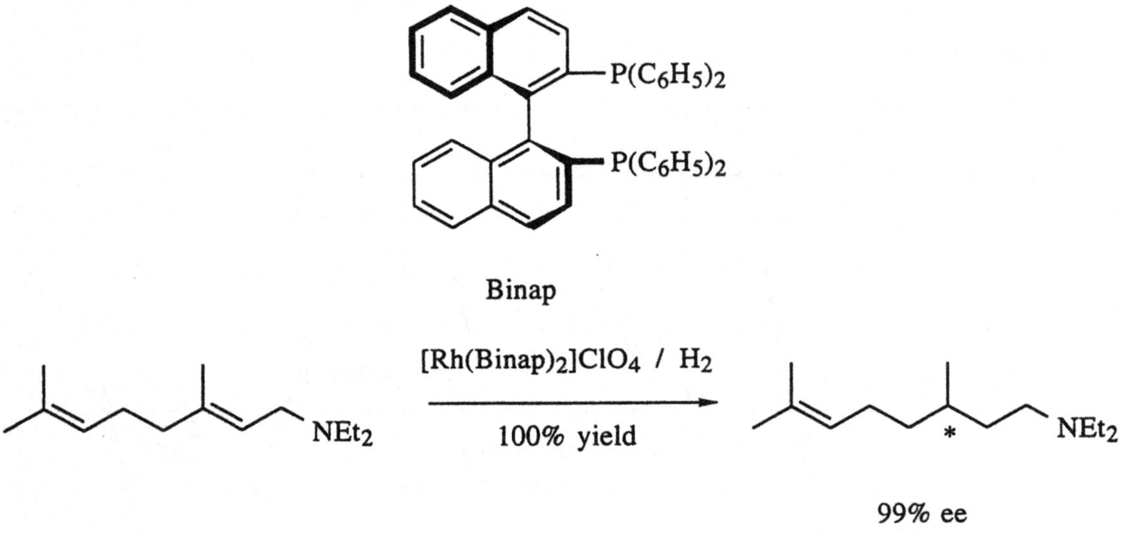

Binap

a)

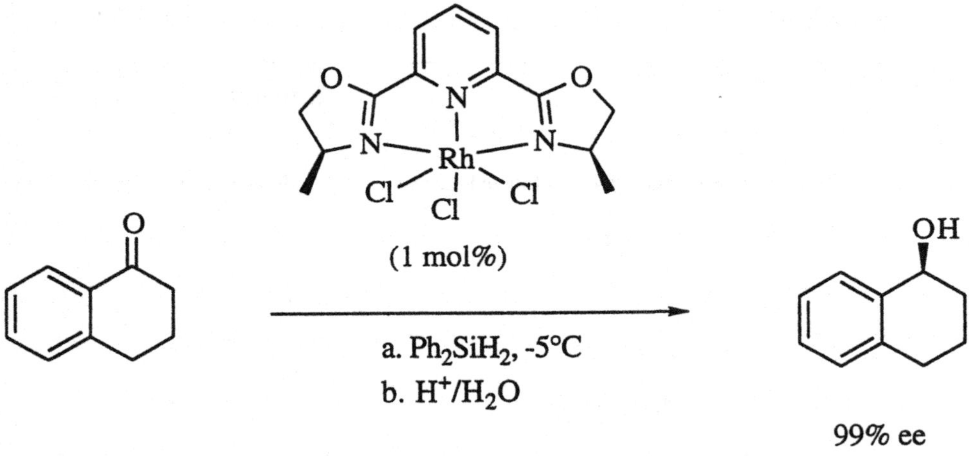

b)

Figure 7.15
Five representative examples of highly enantioselective catalytic reactions (*continued opposite*)

often some irreversible step, such as removal of the kinetically favored product from the reaction system, has to be introduced.

Enantioselective catalysis is an interdisciplinary field, where coordination chemistry has contributed strongly to its development through the design of new ligands and through a judicious choice of metals. Often, however, the exact nature of the catalytic species is not known since it is formed *in situ*. Apart from some

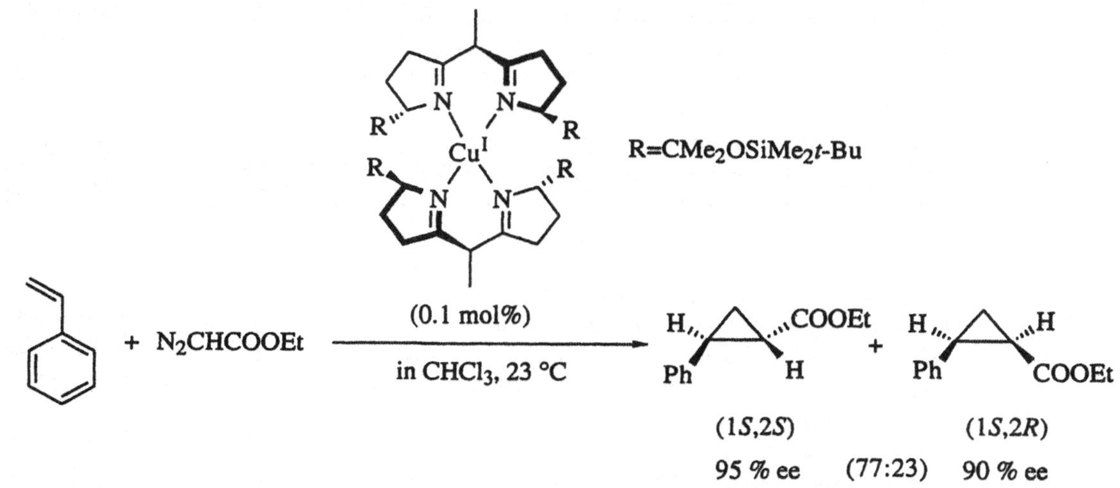
R=CMe$_2$OSiMe$_2$*t*-Bu
+ N$_2$CHCOOEt
(0.1 mol%)
in CHCl$_3$, 23 °C
COOEt
Ph
(1*S*,2*S*)
95 % ee
(77:23)
(1*S*,2*R*)
90 % ee

c)

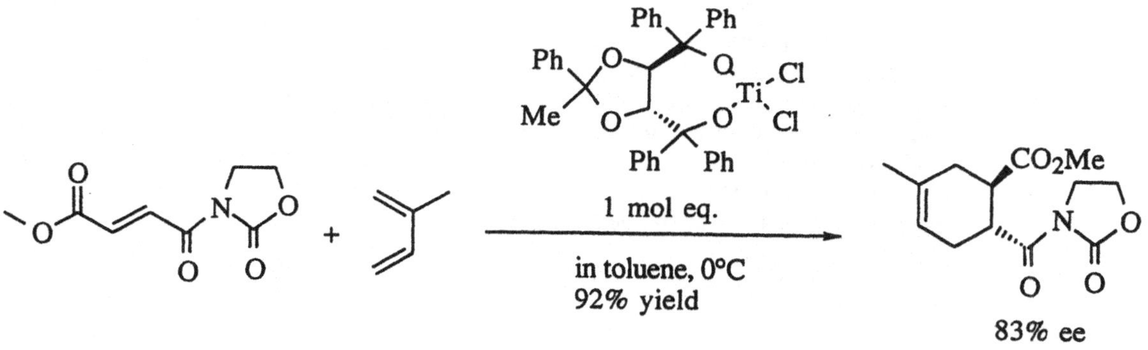
1 mol eq.
in toluene, 0°C
92% yield
CO$_2$Me
83% ee

d)

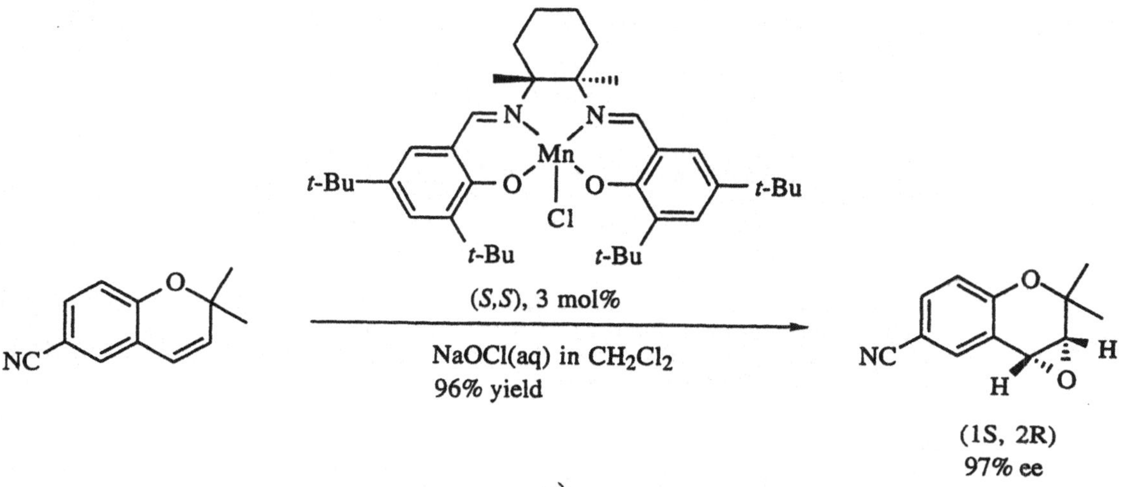
t-Bu
Mn
Cl
(*S*,*S*), 3 mol%
NaOCl(aq) in CH$_2$Cl$_2$
96% yield
NC
(1S, 2R)
97% ee

e)

earlier reports of enantioselective heterogeneous catalysis and polymerization reactions, the first reaction enantioselectively catalyzed by a metal complex was the that shown in Figure 7.14 [96]. The catalyst is a simple Cu^{II} complex with a chiral ligand.

The ee in this reaction was reported to be 6%. Enantioselective catalysis has gone a long way since this first observation was reported. It remains a field in which both rational design and serendipity are important for progress. The former is replacing more and more the latter as knowledge about mechanistic details accumulates. Complicated molecular systems, such as those involved in enantioselective catalysis, will, however, always leave considerable room for human creativity. Brunner, whose contribution to enantioselective catalysis has been very important, has expressed the future possibilities in this field in the following thesis:[†] 'For each and every reaction, where a new stereogenic center is created, exists a transition metal catalyst, which yields only one enantiomer. It just needs to be found.'

One of the central requirements of enantioselective catalysis is a chiral site in the coordination sphere, where the creation of the stereogenic center in the organic molecule from the prochiral precursor takes place. To date it seems that only catalysts which are chiral by virtue of chiral ligands are known to provide this chiral induction. Most often these ligands are bidentate, C_2 symmetric molecules with two phosphorus or nitrogen donors.

Some of the reactions which have been reported are given in Figure 7.15. [97–101]. A common characteristic of all these cases is the C_2 symmetry of the metal complex employed in the catalytic reaction.

Undoubtedly, enantioselective catalysis will develop further and it can therefore be assumed that the number of industrially synthesized chiral compounds will increase significantly in the near future.

7.7   References

1. Wilkins, R. G., *Kinetics and Mechanism of Reactions of Transition Metal Complexes*, 2nd edn, VCH, Weinheim, 1991.
2. Eaton, D. R., Phillips, W. D. and Caldwell, D. J. (1963), *J. Am. Chem. Soc.*, **85**, 397–406.
3. Holm, R. H. and O'Connor, M. J. (1971), *Prog. Inorg. Chem.*, **14**, 241–401.
4. Chernyaev, I. I. (1926), *Izv. Inst. Izuch. Plat. Drugikh Blagorodn. Metal*, **4**, 243–275.
5. Kauffman, G. (1977), *J. Chem. Educ.*, **54**, 86–89.
6. Basolo, F. and Pearson, R. G. (1962), *Prog. Inorg. Chem.*, **4**, 381–453.
7. Basolo, F. and Hammaker, G. S. (1962), *Inorg. Chem.*, **1**, 1–5.
8. Merbach, A. E. (1987), *Pure Appl. Chem.*, **59**, 161–172.
9. Jackson, W. G. and Sargeson, A. M., in *Rearrangements in Ground and Excited States*, P. de Mayo (Ed.), Vol. 2, Organic Chemistry: A Series of Monographs, Academic Press, New York, 1980, pp. 273–378.
10. Jackson, W. G., in *Stereochemistry of Organometallic and Inorganic Compounds*, I. Bernal (Ed.), Vol. 1, Elsevier, Amsterdam, 1986, pp. 255–357.
11. Gielen, M., Dehouck, C., Mokhtan-Jamai, H. and Topart, J. (1972), *Rev. Silicon Germanium Tin Lead Compd*, **1**, 9–33.

[†]Communicated by H. Brunner at the Convention Intercantonale Intercantonale Romande pour l'Enseignement du 3e Cycle en Chimie, in Champéry, Switzerland, 1993.

12. King, R. B. (1991), *J. Math. Chem.*, **7**, 51–68.
13. Sokolov, V. I., *Introduction to Theoretical Stereochemistry* (translated from Russian by N. F. Standen), Gordon and Breach, New York, 1991.
14. Rodger, A. and Schipper, P. E. (1988), *Inorg. Chem.*, **27**, 458–466.
15. Muetterties, E. L. (1972), *Inorg. Chem., Ser. One*, **9**, 37–85.
16. Muetterties, E. L. (1970), *Acc. Chem. Res.*, **3**, 266–273.
17. Von Zelewsky, A. (1981), *Inorg. Chem.*, **20**, 4448–4449.
18. Daul, C., Deiss, E., Gex, J. N., Perret, D., Schaller, D. and Von Zelewsky, A. (1983), *J. Am. Chem. Soc.*, **105**, 7556–7563.
19. Schaller, D. and Von Zelewsky, A. (1979), *J. Chem. Soc., Chem. Commun.*, 948.
20. Abel, E. W., Dimitrov, V. S., Long, N. J., Orrell, K. G., Osborne, A. G., Pain, H. M., Sik, V., Hursthouse, M. B. and Mazid, M. A. (1993), *J. Chem. Soc., Dalton Trans.*, 597–603.
21. Abel, E. W., Dimitrov, V. S., Long, N. J., Orrell, K. G., Osborne, A. G., Sik, V., Hursthouse, M. B. and Mazid, M. A. (1993), *J. Chem. Soc., Dalton Trans.*, 291–298.
22. Abel, E. W., Orrell, K. G., Osborne, A. G., Pain, H. M. and Sik, V. (1994), *J. Chem. Soc., Dalton Trans.*, 111–116.
23. Anderson, P. A., Keene, F. R., Horn, E. and Tiekink, E. R. T. (1990), *Appl. Organomet. Chem.*, **4**, 523–533.
24. Ruch, E. (1977), *Angew. Chem., Int. Ed. Engl.*, **16**, 65–72.
25. Rodger, A. and Johnson, B. F. G. (1988), *Inorg. Chem.*, **27**, 3061–3062.
26. Ray, P. and Dutt, N. K. (1943), *J. Indian Chem. Soc.*, **20**, 81–92.
27. Bailar, J. C., Jr (1958), *Chem. Coord. Compds., Symp., Rome, 1957*, 165–175.
28. Searle, G. H. and Keene, F. R. (1972), *Inorg. Chem.*, **11**, 1006–1111.
29. Fortman, J. J. and Sievers, R. E. (1971), *Coord. Chem. Rev.*, **6**, 331–375.
30. Serpone, N. and Bickley, D. G. (1972), *Prog. Inorg. Chem.*, **17**, 391–566.
31. Gordon, J. G. and Holm, R. H. (1970), *J. Am. Chem. Soc.*, **92**, 5319–5332.
32. Gordon, J. G., OConnor, M. J. and Holm, R. H. (1971), *Inorg. Chim. Acta*, **5**, 381–388.
33. Archer, R. D. (1969), *Coord. Chem. Rev.*, **4**, 243–272.
34. Basolo, F. (1967), *Adv. Chem. Ser.*, **62**, 408–429.
35. Basolo, F. and Pearson, R. G., *Mechanisms of Inorganic Reactions–A Study of Metal Complexes in Solution*, 2nd edn, Wiley, New York, 1967, p. 273.
36. Dasgupta, T. P. (1974), *Inorg. Chem., Ser. Two*, **9**, 63–91.
37. Langford, C. H. and Gray, H. B., *Ligand Substitution Processes*, Benjamin, New York, 1965, Chapter 2.
38. Langford, C. H. and Parris, M. (1972), *Comp. Chem. Kinet.*, **7**, 1–55.
39. Langford, C. H. and Sastri, V. S. (1972), *Inorg. Chem., Ser. One*, 203–267.
40. Langford, C. H. and Stengle, T. R. (1968), *Annu. Rev. Phys. Chem.*, **19**, 193–214.
41. Poon, C. K. (1970), *Inorg. Chim. Acta, Rev.*, **4**, 123–144.
42. Tobe, M. L., in *Studies on Chemical Structure and Reactivity*, J. H. Ridd (Ed.), Methuen, London, 1966, p. 215.
43. Tobe, M. L., *Inorganic Reaction Mechanisms–Studies in Modern Chemistry*, Nelson, London, 1972.
44. Barraclough, C. G., Boschem, R. W., Fee, W. W., Jackson, W. G. and McTigue, P. T. (1971), *Inorg. Chem.*, **10**, 1994–1997.
45. Werner, A., *Neuere Anschauungen auf dem Gebiete der Anorganischen Chemie*, 3. Auflage, Vieweg, Braunschweig, 1913.
46. Garbett, K. and Gillard, R. D. (1965), *J. Chem. Soc.*, 6084–6100.
47. Garbett, K. and Gillard, R. D. (1966), *J. Chem. Soc. A*, 204–206.
48. Garbett, R. D., Gillard, R. D. and Staples, P. J. (1966), *J. Chem. Soc. A*, 201–204.
49. Hua, X., *Chiral Building Blocks Ru(L^L)₂ for Coordination Compounds*, Diss. No. 1047, University of Fribourg, Fribourg, 1993.
50. Hua, X. and Von Zelewsky, A. (1991), *Inorg. Chem.*, **30**, 3796–3798.
51. Jonassen, H. B., Bailar, J. C., Jr, and Huffman, E. H. (1948), *J. Am. Chem. Soc.*, **70**, 756–758.
52. Kirschner, S., Wei, Y.-K. and Bailar, J. C., Jr (1957), *J. Am. Chem. Soc.*, **79**, 5877–5880.

53. Liu, C. F., Liu, N. C. and Bailar, J. C., Jr (1964), *Inorg. Chem.*, **3**, 1085–1087.
54. Burstall, F. H. (1936), *J. Chem. Soc.*, 173–175.
55. Burstall, F. H., Dwyer, F. B. and Gyarfas, E. C. (1950), *J. Chem. Soc.*, 953–955.
56. Bailar, J. C., Jr, and Auten, R. W. (1934), *J. Am. Chem. Soc.*, **56**, 774–776.
57. Boucher, L. J., Kyuno, E. S. and Bailar, J. C., Jr (1964), *J. Am. Chem. Soc.*, **86**, 3656–3660.
58. Dang, L. X. and Pettitt, B. M. (1987), *J. Chem. Phys.*, **86**, 6560–6561.
59. Pettitt, B. M. and Rossky, P. J. (1986), *J. Chem. Phys.*, **84**, 5836–5844.
60. Zhong, E. C. and Friedman, H. L. (1988), *J. Phys. Chem.*, **92**, 1685–1692.
61. Kyuno, E. and Bailar, J. C., Jr (1966), *J. Am. Chem. Soc.*, **88**, 5447–5451.
62. Kyuno, E. and Bailar, J. C., Jr (1966), *J. Am. Chem. Soc.*, **88**, 1120–1124.
63. Kyuno, E. and Bailar, J. C., Jr (1966), *J. Am. Chem. Soc.*, **88**, 1125–1128.
64. Kyuno, E., Boucher, L. J. and Bailar, J. C., Jr (1965), *J. Am. Chem. Soc.*, **87**, 4458–4462.
65. Forster, D. (1979), *J. Chem. Soc., Dalton Trans.*, 1639–1645.
66. Stille, J. K. and Lau, K. S. Y. (1977), *Acc. Chem. Res.*, **10**, 434–442.
67. Basolo, F., Bailar, J. C., Jr, and Tarr, R. B. (1950), *J. Am. Chem. Soc.*, **72**, 2433–2438.
68. Heneghan, L. F. and Bailar, J. C., Jr (1953), *J. Am. Chem. Soc.*, **75**, 1840–1841.
69. Chassot, L. and Von Zelewsky, A. (1986), *Helv. Chim. Acta*, **69**, 1855–1857.
70. Sandrini, D., Maestri, M., Balzani, V., Chassot, L. and Von Zelewsky, A. (1987), *J. Am. Chem. Soc.*, **109**, 7720–7724.
71. Von Zelewsky, A., Suckling, A. P. and Stoeckli-Evans, H. (1993), *Inorg. Chem.*, **32**, 4585–4593.
72. Lehn, J.-M. (1988), *Angew. Chem., Int. Ed. Engl.*, **27**, 89–112.
73. Balzani, V. and De Cola, L. (Eds), *Supramolecular Chemistry, NATO ASI Series*, Kluwer, Dordrecht, 1992.
74. Bernal, I., Cai, J. and Myrczek, J. (1993), *Polyhedron*, **12**, 1157–1162.
75. Yoneda, H. and Miyoshi, K., in *Coordination Chemistry. A Century of Progress*, G. B. Kauffman (Ed.), ACS Symposium Series, No. 565, 1994, pp. 308–317.
76. Barton, J. K. (1986), *Science*, **233**, 727–734.
77. Hiort, C., Lincoln, P. and Nordén, B. (1993), *J. Am. Chem. Soc.*, **115**, 3448–3454.
78. Buckingham, D. A., Marzilli, L. G. and Sargeson, A. M. (1967), *J. Am. Chem. Soc.*, **89**, 825–830.
79. Keene, F. R., Ridd, M. J. and Snow, M. R. (1983), *J. Am. Chem. Soc.*, **105**, 7075–7081.
80. Lindoy, L. F. and Bush, D. H. (1974), *Inorg. Chem.*, **13**, 2494–2498.
81. McMurry, T. J., Raymond, K. N. and Smith, P. H. (1989), *Science*, **244**, 938–943.
82. Taube, H., *Electron Transfer Reactions of Complex Ions in Solutions, Current Chemical Concepts.–A Series of Monographs*, Academic Press, New York, 1970.
83. Marcus, R. A. (1964), *Annu. Rev. Phys. Chem.*, **15**, 155–196.
84. Sutin, N. (1982), *Acc. Chem. Res.*, **15**, 275–282.
85. Bernauer, K. (1991), *Met. Ions Biol. Syst.*, **27**, 265–289.
86. Lappin, A. G. and Marusak, R. A. (1991), *Coord. Chem. Rev.*, **109**, 125–180.
87. Lappin, A. G, *Redox Mechanisms in Inorganic Chemistry*, Ellis Horwood, Chichester, 1994.
88. Geselowitz, D. A. and Taube, H. (1980), *J. Am. Chem. Soc.*, **102**, 4525–4526.
89. Bernauer, K., Fuchs, E. and Hugi-Cleary, D. (1994), *Inorg. Chim. Acta*, **218**, 73–79.
90. Bernauer, K., Pousaz, P., Porret, J. and Jeanguenat, A. (1988), *Helv. Chim. Acta*, **71**, 1339–1348.
91. Stinson, S. C. (1992), *Chem. Eng. News*, **70**, 46–76.
92. Brunner, H. and Zettlmeier, W., *Handbook of Enantioselective Catalysis with Transition Metal Complexes–Ligands–References*, Vol. II, VCH, Weinheim, 1993.
93. Brunner, H. and Zettlmeier, W., *Handbook of Enantioselective Catalysis with Transition Metal Complexes–Products and Catalysts*, Vol. I, VCH, Weinheim, 1993.
94. Noyori, R., *Asymmetric Catalysis in Organic Synthesis*, Wiley, New York, 1994.
95. Gold, V., Loening, K. L., McNaught, A. D. and Sehmi, P., *Compendium of Chemical Terminology–IUPAC Recommendations*, Blackwell, Oxford, 1987.

96. Nozaki, H., Takaya, H., Moriuti, S. and Noyori, R. (1968), *Tetrahedron*, **24**, 3655–3669.
97. Jacobsen, E. N., Zhang, W., Muci, A. R., Ecker, J. R. and Li, D. (1991), *J. Am. Chem. Soc.*, **113**, 7063–7064.
98. Müller, D., Umbricht, G., Weber, B. and Pfaltz, A. (1991), *Helv. Chim. Acta*, **74**, 232–240.
99. Nishiyama, H., Itoh, Y., Matsumoto, H., Park, S. B. and Itoh, K. (1994), *J. Am. Chem. Soc.*, **116**, 2223–2224.
100. Noyori, R. (1990), *Science*, **248**, 1194–1199.
101. Noyori, R. (1990), *Science*, **249**, 844.

Appendix I

Overview of Point Groups

1 The Groups C_1, $C_s = C_h$, $C_i = S_2$

C_1	E	
C_s	E	σ_h
C_i	E	i

2 The Groups C_n ($n = 2, 3, \ldots, 8$)

C_2	E	C_2						
C_3	E	C_3	C_3^2					
C_4	E	C_4	C_2	C_4^3				
C_5	E	C_5	C_5^2	C_5^3	C_5^4			
C_6	E	C_6	C_3	C_2	C_3^2	C_6^5		
C_7	E	C_7	C_7^2	C_7^3	C_7^4	C_7^5	C_7^6	
C_8	E	C_8	C_4	C_2	C_4^3	C_8^3	C_8^5	C_8^7

3 The Groups D_n ($n = 2, 3, 4, 5, 6$)

D_2	E	$C_2(z)$	$C_2(y)$	$C_2(x)$		
D_3	E	$2C_3$	$3C_2$			
D_4	E	$2C_4$	C_2	$2C'_2$	$2C''_2$	
D_5	E	$2C_5$	$2C_5^2$	$5C_2$		
D_6	E	$2C_6$	$2C_3$	C_2	$3C'_2$	$3C''_2$

4 The Groups C_{nv} ($n = 2, 3, 4, 5, 6$)

C_{2v}	E	C_2	$\sigma_v(xz)$	$\sigma_v'(yz)$					
C_{3v}	E	$2C_3$	$3\sigma_v$						
C_{4v}	E	$2C_4$	C_2	$2\sigma_v$	$2\sigma_d$				
C_{5v}	E	$2C_5$	$2C_5^2$	$5\sigma_v$					
C_{6v}	E	$2C_6$	$2C_3$	C_2	$3\sigma_v$	$3\sigma_d$			

5 The Groups C_{nh} ($n = 2, 3, 4, 5, 6$)

C_{2h}	E	C_2	i	σ_h							
C_{3h}	E	C_3	C_3^2	σ_h	S_3	S_3^5					
C_{4h}	E	C_4	C_2	C_4^3	i	S_4^3	σ_h	S_4			
C_{5h}	E	C_5	C_5^3	C_5^4	σ_h	S_5	S_5^7	S_5^3	S_5^9		
C_{6h}	E	C_6	C_3	C_2	C_3^2	C_6^5	i	S_3^5	S_6^5	σ_h	S_6	S_3

6 The Groups D_{nh} ($n = 2, 3, 4, 5, 6$)

D_{2h}	E	$C_2(z)$	$C_2(y)$	$C_2(x)$	i	$\sigma(xy)$	$\sigma(xz)$	$\sigma(yz)$				
D_{3h}	E	$2C_3$	$3C_2$	σ_h	$2S_3$	$3\sigma_v$						
D_{4h}	E	$2C_4$	C_2	$2C_2'$	$2C_2''$	i	$2S_4$	σ_h	$2\sigma_v$	$2\sigma_d$		
D_{5h}	E	$2C_5$	$2C_5^2$	$5C_2$	σ_h	$2S_5$	$2S_5^3$	$5\sigma_v$				
D_{6h}	E	$2C_6$	$2C_3$	C_2	$3C_2'$	$3C_2''$	i	$2S_3$	$2S_6$	σ_h	$3\sigma_d$	$3\sigma_v$

7 The Groups D_{nd} ($n = 2, 3, 4, 5, 6$)

D_{2d}	E	$2S_4$	C_2	$2C_2'$	$2\sigma_d$				
D_{3d}	E	$2C_3$	$3C_2$	i	$2S_6$	$3\sigma_d$			
D_{4d}	E	$2S_8$	$2C_4$	$2S_8^3$	C_2	$4C_2'$	$4\sigma_e$		
D_{5d}	E	$2C_5$	$2C_5^2$	$5C_2$	i	$2S_{10}^3$	$2S_{10}$	$5\sigma_d$	
D_{6d}	E	$2S_{12}$	$2C_6$	$2S_4$	$2C_3$	$2S_{12}^5$	C_2	$6C_2'$	$6\sigma_d$

8   The Groups S_n ($n = 4, 6, 8$)

S_4	E	S_4	C_2	S_4^3			
S_6	E	C_3	C_3^2	i	S_6^5	S_6	
S_8	E	S_8	C_4	S_8^3	C_2	S_8^5	C_4^3	S_8^7

9   The Cubic Groups

T	E	$4C_3$	$4C_3^2$	$3C_2$					
T_d	E	$8C_3$	$3C_2$	$6S_4$	$6\sigma_d$				
T_h	E	$4C_3$	$4C_3^2$	$3C_2$	i	$4S_6$	$4S_6^2$	$3\sigma_d$	
O	E	$8C_3$	$3C_2$	$6C_4$	$6C_2'$				
O_h	E	$8C_3$	$6C_2$	$6C_4$	$3C_2$	i	$6S_4$	$8S_6$	$3\sigma_h$	$6\sigma_d$

10   The Groups I, I_h

I	E	$12C_5$	$12C_5^2$	$20C_3$	$15C_2$					
I_h	E	$12C_5$	$12C_5^2$	$20C_3$	$15C_2$	i	$12S_{10}$	$12S_{10}^3$	$20S_6$	15σ

11   The Groups $C_{\infty v}$, $D_{\infty h}$

$C_{\infty v}$	E	C_2	$2C_\infty \phi \ldots$	$\infty\sigma_v$			
$D_{\infty h}$	E	$2C_\infty \phi \ldots$	$\infty\sigma_v$	i	$2S_\infty \phi \ldots$	∞C_2	

Appendix II

Glossary

Absolute configuration. The arrangement of atoms in a pair of enantiomers and its description by a chirality symbol.

Achirotopic. A point in a molecule (generally, but not necessarily, an atomic center) which has an achiral site symmetry.

Allogon. Rarely used term for isomers having different coordination geometry. Synonymous with polytopal isomer.

Alterdentate ligand. A ligand offering two or more symmetrically equivalent coordination sites to a metal.

Ambidentate ligand. A ligand that can coordinate in at least two non-equivalent ways to a metal atom.

Antihelical. A molecular strand that is locally helical, but as a whole achiral through alternating senses of consecutive helical elements.

Asymmetric center (atom). Has been used mostly for T-4 centers that have four different ligands (substituents). In coordination chemistry it has also sometimes been used for chiral OC-6 centers. The usage of the term 'asymmetric center or atom' is now discouraged. It should be replaced by 'chiral center.'

Asymmetric synthesis (catalysis). A synthetic (catalytic) method, which yields unequal amounts of the two enantiomers of a chiral substance, starting from a prochiral precursor. A more appropriate term is 'Enantioselective Synthesis (Catalysis).'

Asymmetric. Devoid of any symmetry.

Bailar inversion. A true inversion that occurs by a substitution reaction at an OC-6 center.

Bailar twist. A rearrangement of a chiral OC-6 $[M(L^\wedge L)_3]$ complex to its enantiomer through an achiral TP-6 intermediate which has D_{3h} symmetry.

Cage ligands. Polycyclic molecule occupying all ligand positions in a complex. All ligand atoms lie in closed cycles.

Catoptromer. Rarely used synonym for enantiomer.

Chelate ligands. Ligands that coordinate through at least two donor atoms to one central metal, thereby forming a chelate ring.

Chiral axis. Axis in a chiral structure in which the ligands responsible for the chirality are bound to several atoms lying on a line.

Chiral center. Center in a chiral structure to which the ligands responsible for the chirality are bound.

Chiral plane. A planar segment of a chiral molecule which shows chirality in 2D space.

Chiral shift reagent. Enantiomerically pure chiral molecule that can cause different chemical shifts for two enantiomers through diastereomeric interactions.

Chiral, chirality. The property of any kind of object, here in particular of molecules, that the object itself, and its realized mirror image, are not congruent.

Chirality descriptors. Symbols that give the absolute configuration of one enantiomer. Three types are used: R/S or C/A are based on the priority rules, they are used for central chirality; Δ/Λ or δ/λ are defined by skew lines; $\overrightarrow{\Delta}/\overrightarrow{\Lambda}$ are defined by the oriented line system.

Chiroptical methods. Optical methods using planar or circularly polarized light.

Chirotopic. A point in a molecule (generally, but not necessarily, an atomic center) which has a chiral site symmetry.

CIP system. Acronym for the 'Cahn–Ingold–Prelog' system. A system for the assignement of priorities to atoms and groups in molecules and the stereochemical descriptors derived from these priorities.

Configuration index. A number (composed of one, two, or three digits) that characterizes the diastereomers of various arrangements of monodentate ligands around a coordination center. It is determined by the priorities of the ligands according to conventional rules.

Configuration number. Synonym for configuration index.

Configuration. An atomic spatial arrangement that is fixed by the chemical bonding in a molecule and that cannot be altered without breaking bonds (contrasted with conformation).

Configurational isomers. Stereoisomers that differ in configuration.

Configurationally inert. A molecular species that can be studied by chemical methods, without changing configuration.

Conformation. A spatial arrangement of atoms that can be altered by (usually rapid) rotation about bonds.

Conglomerate crystals. The macroscopic mixture of two enantiomorphic crystal types (of the same space group), obtained by crystallization of a racemic mixture.

Constitutional isomers. Isomers with different connectivities between the atoms.

Coordination geometry. The idealized polyhedron of a coordination center most closely resembling the real arrangement of a coordination unit.

Coordination isomers. A special type of constitutional isomers in which the metal to ligand bond connectivities are different.

Coordination number. The number of nearest neighbors (number of ligating atoms) of a coordination center.

Coordination symmetry. The site symmetry of a coordination center. Often the coordination geometry is falsely called coordination symmetry.

Desymmetrization. Reduction in symmetry by processes (such as substitution), or by other causes such as a Jahn–Teller distortion.

Diastereomers (diastereoisomers). Stereoisomers that are not enantiomers.

Diastereotopic ligands. Homomorphic ligands (including phantom ligands) in constitutionally equivalent locations that are not related by a symmetry operation. Substitution of one or the other of a pair of diastereotopic ligands by a new ligand will yield diastereomers.

Dissymmetric. Lacking symmetry elements of improper rotation, thus equivalent to chiral, which should preferably be used.

Enantiomer. One of the two of a pair of noncongruent molecular species that are mirror images.

Enantiomeric excess (ee). The purity of an enantiomer with respect to the molecular species with opposite chirality. For definition, see p. 54. Mostly used only for the major component, i.e. $0 \leqslant ee \leqslant 1$. If used for the whole range of compositions, $-1 \leqslant ee \leqslant 1$.

Enantiomerically pure. Limiting case of a substance consisting of only one enantiomer of a chiral pair.

Enantiomorph. The equivalent of enantiomer, but used for macroscopic objects.

Enantiopure. Short form of enantiomerically pure.

Enantiotopic. Homomorphic ligands (including phantom ligands) related by a symmetry element of improper rotation, but not by one of proper rotation. Substitution of one or the other of a pair of enantiotopic ligands yields two enantiomers.

Geometric isomers. Misnomer for diastereomers.

Helix, helical. A geometrical structure in 3D space with defined chirality that can be related to a screw axis. The minimum requirement is for there to be two skew lines in space.

Hetereotopic ligands. Homomorphic ligands (including phantom ligands) that are either enantiotopic or diastereotopic.

Heteroconversion. Isomerization that leads to a diastereomer.

Heteroleptic center. A coordination center with more than one kind of ligand.

Heteromorphic ligands. Ligands that differ when detached from the coordination center.

Homochiral. Two or more elements of equal chirality in a molecular species.

Homoconversion. Isomerization (rearrangement) that yields either the other of a pair of enantiomers or a species with identical configuration.

Homoleptic coordination center. A coordination center where all ligands are homomorphic.

Homomorphic ligands. Ligands that are identical when detached from the coordination center.

Homotopic ligands. A pair of homomorphic ligands related by a proper rotation symmetry element. Homotopic ligands (including phantom ligands) yield identical products when either of them are substituted by a new ligand.

Hydrate isomers. Coordination compounds of identical stoichiometric composition, having water molecules either coordinated to a central metal or as so-called crystal water.

Improper rotation. A symmetry operation that consists in rotation around an axis and mirroring on a plane perpendicular to that axis. A mirror plane and a center of inversion are special cases of improper rotation symmetry elements (S_1 and S_2, respectively).

Inversion of configuration. A rearrangement where a molecule changes its absolute configuration. Inversion of configurations described by the skew line or by the oriented line reference systems are always true inversions, whereas those described by R/S and C/A central chirality symbols may be either true or pseudo inversions.

Ionization isomers. Isomerism in a coordination compound due to the presence of an ionic species (in general anionic) either as ligand of a central metal or as constituent of an ionic lattice.

Isoexchange. An intramolecular process, such as a pseudo-rotation, leading to a stereochemically identical arrangement.

Isomers. Compounds or species of identical stoichiometric composition but different structure. Molecular species that are isomers (including complex ions) must have the same mass.

Jahn–Teller instability. The inherent instability of a high-symmetry arrangement of atoms in a molecule due to the fact that this symmetry causes a degeneracy of the ground state. This generally leads to static or dynamic desymmetrizations.

Ligand. An ionic or neutral, monoatomic or polyatomic species that is capable of binding to another center. In coordination chemistry, the central atom is a metallic element in a given oxidation state and the ligands are Lewis bases.

Ligating atom. The atom(s) that is (are) bound or can bind to a coordination center.

Linkage isomers. Isomeric species in which a ligand is coordinated to the central metal through either of two distinguishable ligating atoms.

Macrocyclic ligands. Ligands with a cyclic structure that comprise a ring that has generally >6 atoms and at least three ligating atoms.

Metric stereochemistry. That part of stereochemistry in which primarily quantitative aspects, like the values of internal coordinates of molecules are considered.

Monodentate. A ligand coordinated to one center through only one ligating atom.

Octopus ligand. A ligand structure that combines the properties of a polydentate with that of a chelate ligand, thereby forming open-cage complexes.

Optical activity. A solution property of a substance to rotate the plane of polarized light.

Optical isomers. Misnomer for enantiomers.

Optical rotatory dispersion (ORD). Optical activity given as a function of wavelength.

Oriented-line reference system. A reference system for chiral objects defined by two non-intersecting vectors in 3D space that are not parallel.

Oxidative addition. A reaction type where both the coordination number and the oxidation number of the central metal atom increases. Most often SP-4 d^8 to OC-6 d^6.

Phantom ligand. A stereochemically active lone pair. Considered to be massless for the determination of the ligand priorities.

Polarimetry. The measurement of the optical activity by plane polarized light.

Polyhedral symbol. The designation of ligating atom arrangements around a coordination center according to idealized polyhedra by symbols conventionally fixed by IUPAC.

Polytopal isomerism. Isomerism of coordination species due to different coordination polyhedra of the same coordination number.

Priority number. The relative ranking of a ligand atom on the scale given by the CIP priority rules.

Prochirality. The property of a molecular species that possesses enantiotopic ligands (including phantom ligands) to yield a chiral molecule upon a substitution of one of a pair of enantiotopic ligands.

Proper rotation. A symmetry operation that leads to congruent arrangements of atoms in a molecule (or generally of all the elements of an object) by a simple rotation.

Prostereoisomeric. Species having heterotopic ligands are prostereoisomeric, giving either enantiomers (in the case of enantiotopic ligands) or diastereomers (in the case of diastereotopic ligands), if either one of a pair of such ligands is substituted.

Pseudo-asymmetric center. A T-4 center with four different ligands that is not a center of chirality. Pseudo-asymmetry occurs also for other coordination numbers, e.g. OC-6).

Pseudo-inversion. A change from R to S or C to A or vice versa, that comes about solely by the substitution of one single monoatomic ligand through changes in CIP priorities.

Racemate. General description for a substance occurring as a $1:1$ enantiomeric mixture of chiral constituents.

Racemic compound. A racemate in which the Gibbs free energy of the pure racemate is lower than that of the pure enantiomeric constituents.

Racemic mixture. A racemate in which the Gibbs free energy of the pure racemate is higher than that of the pure enantiomeric constituents.

Racemic modification. Synonym for racemic compound.

Racemic solid solution. A crystallographically ordered solid composed of a racemate that shows a statistical distribution of the two enantiomeric constituents.

Racemization. The thermodynamically allowed process (in an achiral environment) yielding finally a racemate from one of the enantiomers.

Ray–Dutt twist. A rearrangement of a chiral OC-6 $[M(L\char`\^L)_3]$ complex to its enantiomer through an achiral TP-6 intermediate which has C_{2v} symmetry.

Reductive elimination. Simultaneous reduction of coordination number and oxidation number of a coordination center in a reaction sequence. Most often OC6 d^6 to SP-4 d^8.

Seniority numbers. Misnomer for priority numbers.

Site symmetry symbols. Misleading synonym for polyhedral symbol.

Site symmetry. The symmetry given by the symmetry elements of the site under consideration.

Skew-line reference system. The reference system for the designation of helical chirality consisting of two skew lines in 3D space. See also oriented line reference system.

Steering wheel reference system. Reference system for descriptors of central chirality.

Stereo centers. Graphic symbols for the representation of coordination polyhedra, indicating the center to ligand connectivity.

Stereo pair. Two 2D images of an object from that a 3D picture that can be visionally reconstructed.

Stereo vision. The mental reconstruction of a 3D object from two 2D images, either by looking through a pair of lenses or unaided.

Stereochemically active electron pair. An electron pair that manifests its presence by occupying a ligating 'atom' site.

Stereogenic center. A center where the interchange of two heteromorphic ligands leads to a stereoisomer. If the center is chirotopic, the stereoisomers are enantiomers.

Stereognostic. A property of a ligands to recognize the particular coordinating properties of an oxo ion, especially with respect to form complexes of high stability.

Stereoisomerism. Any kind of isomers (enantiomers or diastereomers) that are different only in the spatial arrangement of the atoms.

Structural isomers. Synonym for constitutional isomers. Its use is discouraged.

Substitutionally inert. A coordination center where substitution reactions are generally sufficiently slow as to allow for the isolation of complexes that are not necessarily thermodynamically the most stable.

Symmetry element. A geometrical object (point in space, axis, plane) that is invariant under a given symmetry operation. Two general types exist: proper and improper rotation axes. The improper rotation axis of order one is equivalent to a symmetry plane and that of order two to a center of inversion.

Symmetry operation. The operation carried out on an object that leads to a congruent representation of the object.

Symmetry site term. Misnomer for polyhedral symbol.

Topographic stereochemistry. Qualitative considerations in stereochemistry, such as number and kinds of isomers and the topography of rearrangements. As opposed to metric stereochemistry.

Topological isomers. Isomers that can be transformed into each other only by a temporary rearrangement of connectivities. The two enantiomers of a threefoil knot are topological isomers.

Topological stereochemistry. That part of topographical stereochemistry that considers such molecular properties which change only if connectivities are rearranged.

Trans influence series. The series established by Chernyaev that gives the ligand in order of their directing influence in substitution reactions at SP-4 centers (usually Pt^{II}).

Vicinal elements. Dubious term for ligands that are chiral by themselves.

Appendix III

Ligands with various types of donor atoms

1. N-ligands

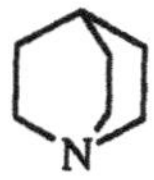

quin

en

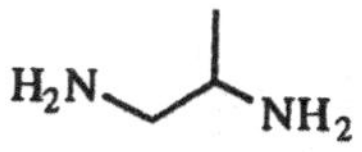

pn

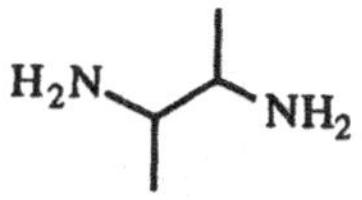

bn

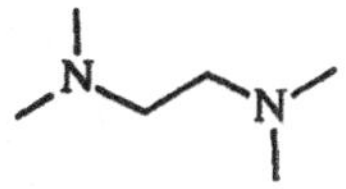

tmen

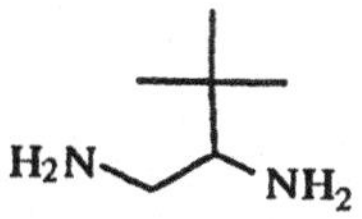

dmbn

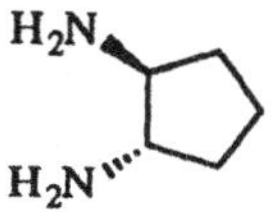

cptn

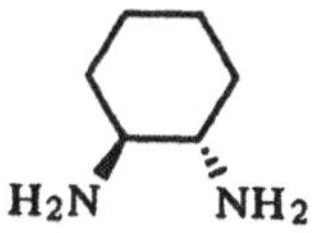

dach

tn

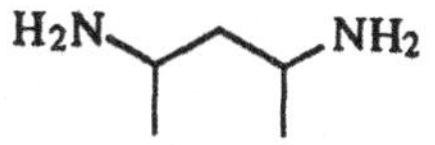

ptn

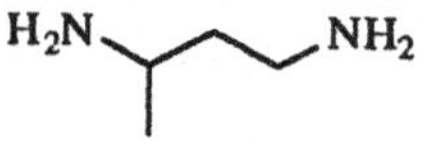

1,3-bn

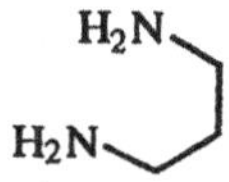

1,3-pn

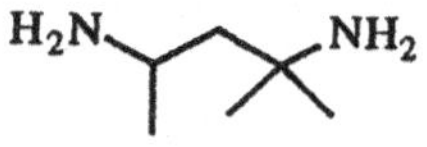

mptn

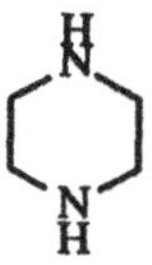

pip

daco

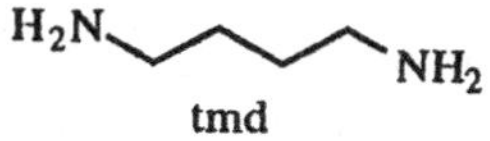

tmd

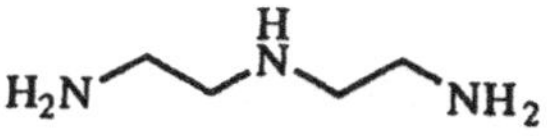

dien

dptn

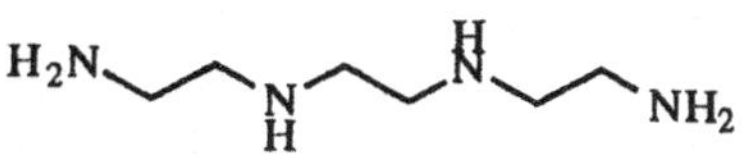

trien

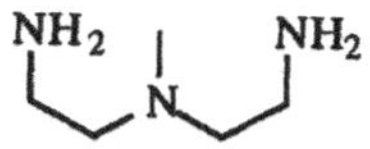

dema

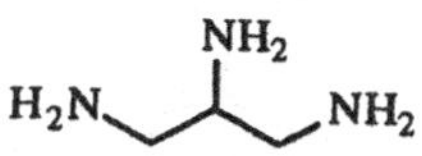

1,2,3-pn

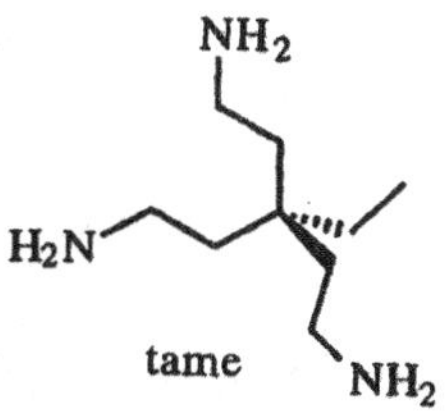

tame

2,3,2-tet

2,2,3-tet

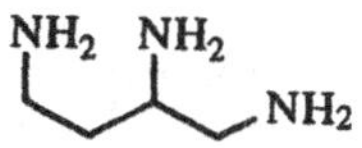

trab

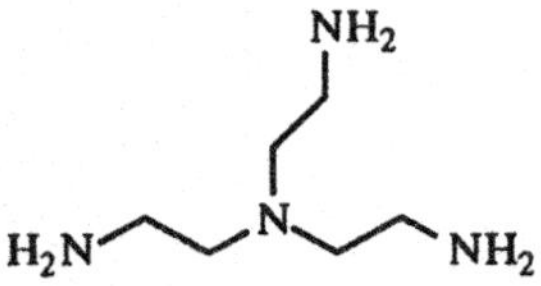

tren

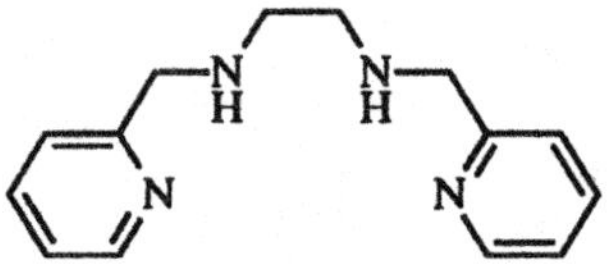

picpn

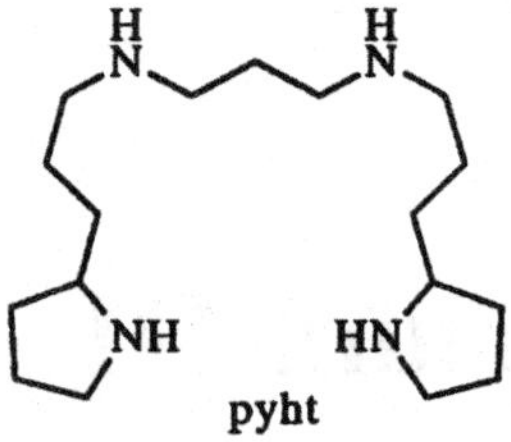

pyht

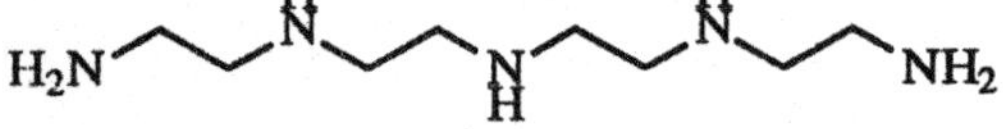

tetraen

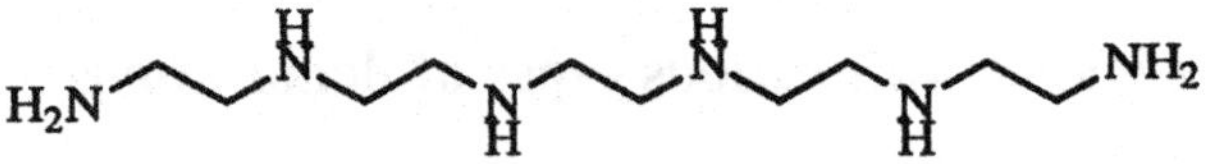

linpen
pentaen

imd

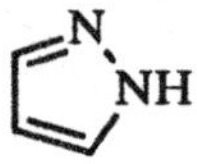

pyz

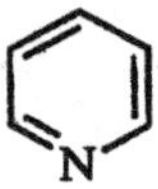

py

pydz

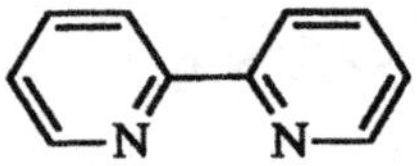

bpy

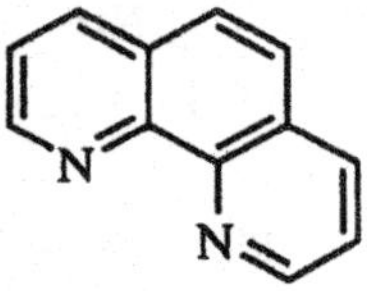

phen

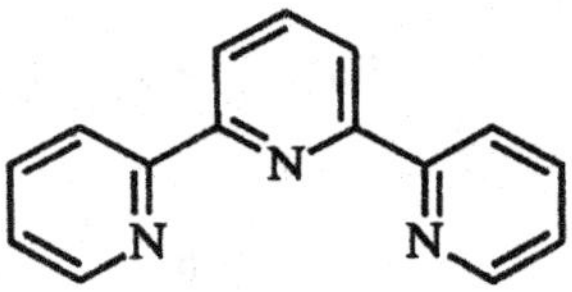

terpy

bpym

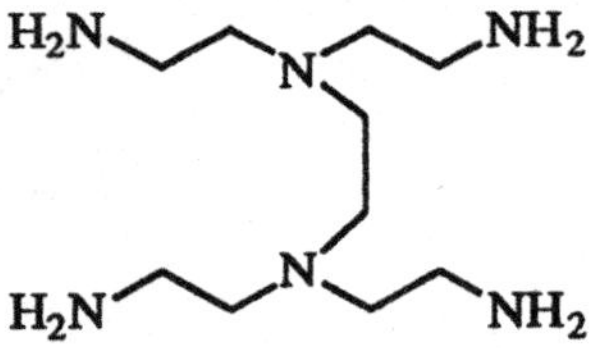

penten

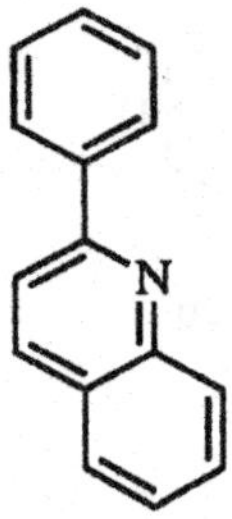

thq

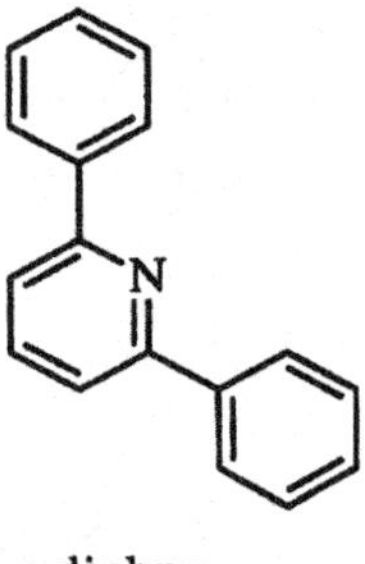

diphpy

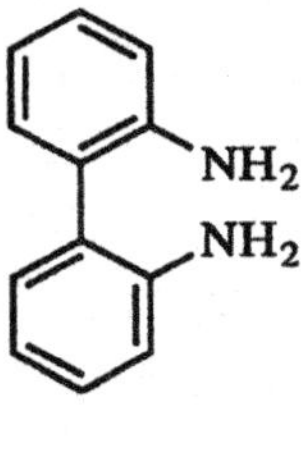

2,2'-diaminobiphenyl

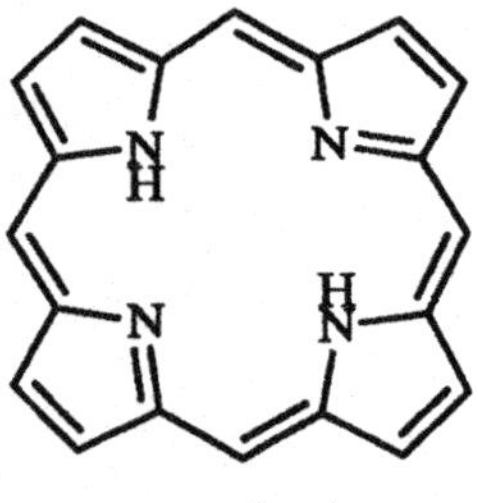

porphyrine

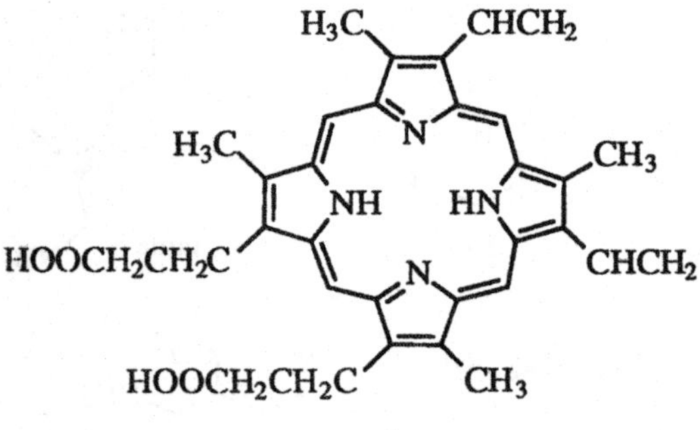

heme

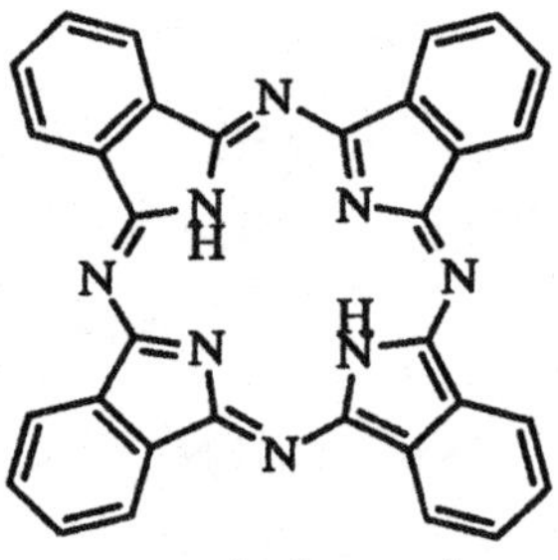

phtalocyanine

2. N/O-ligands

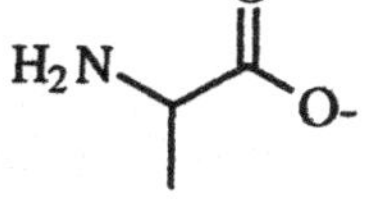

alaO

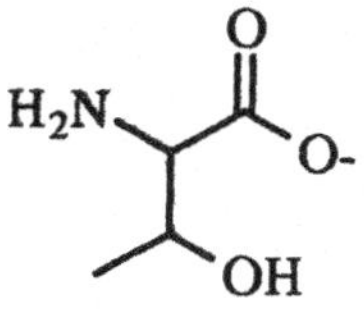

thrO

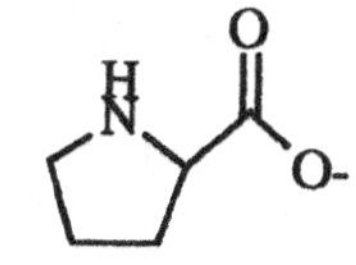

proO

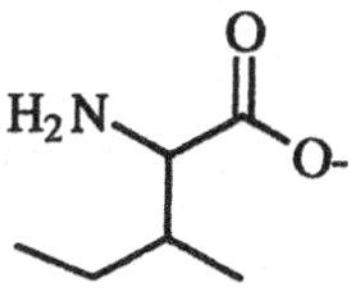

ileuO

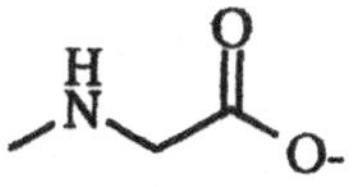

sarcosinato

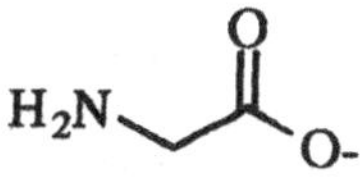

glyO

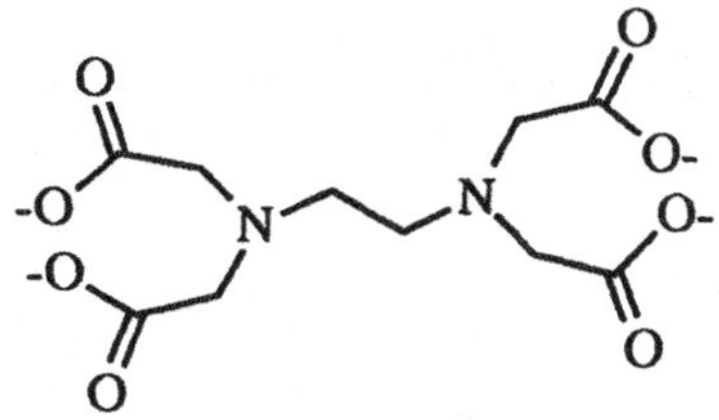

edta

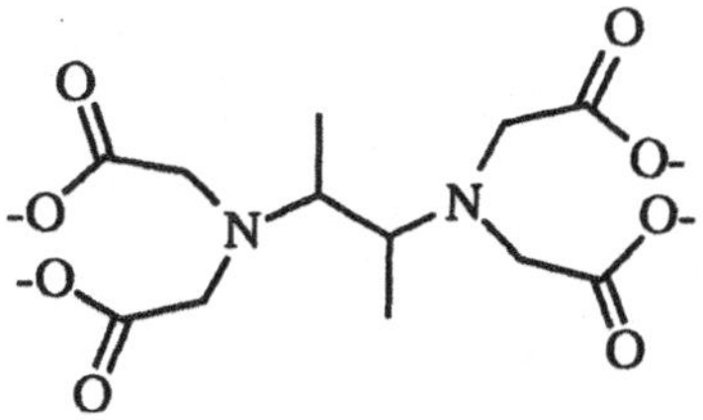

dbta

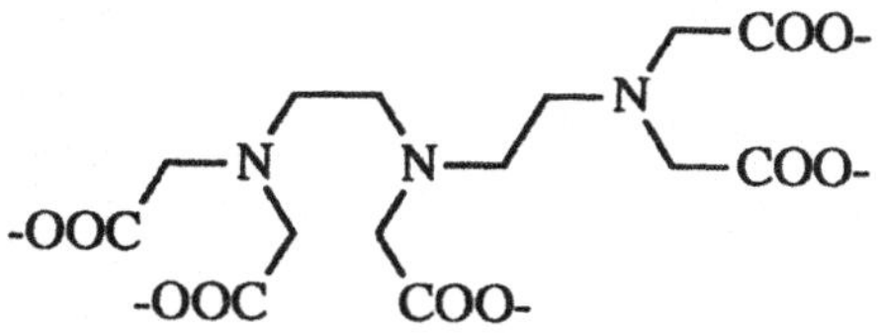

dtpa

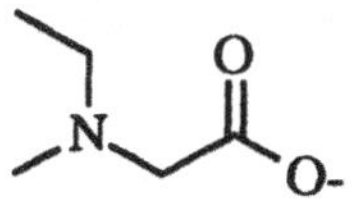

AMAC

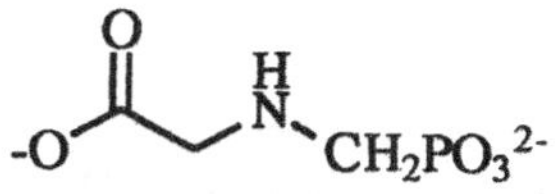

PMG

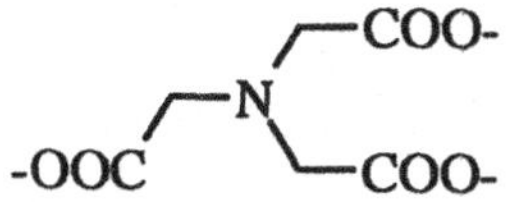

nta

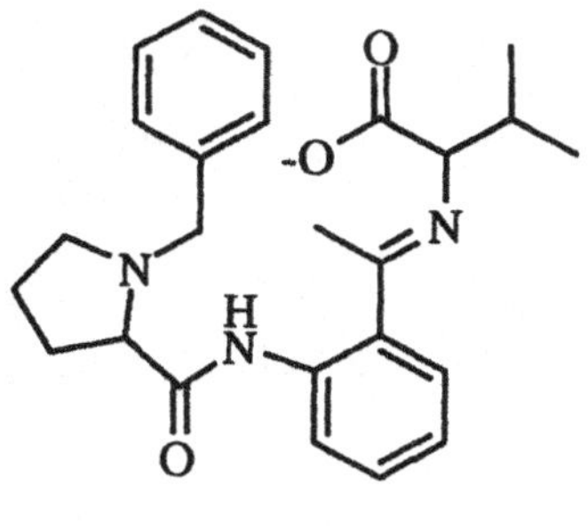

BPAAP-valO

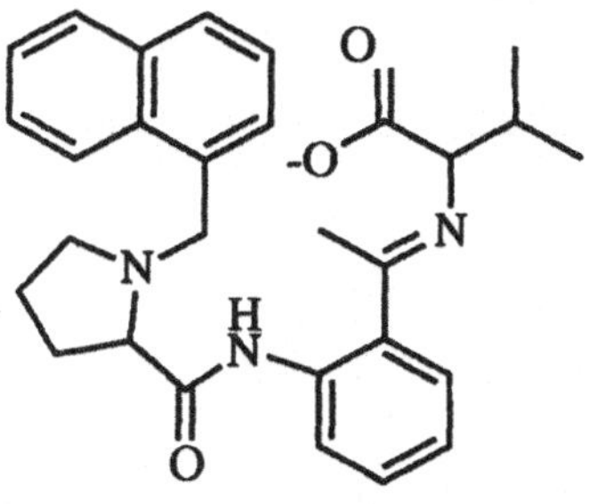

NPAAP-valO

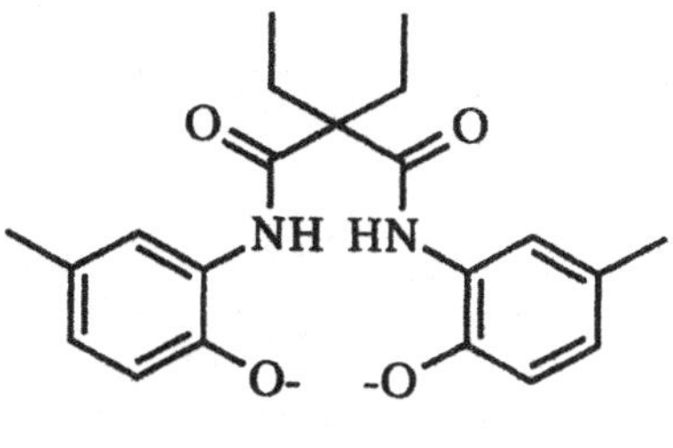

depa

3. O/S-ligands

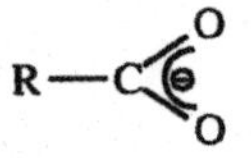

carboxylate 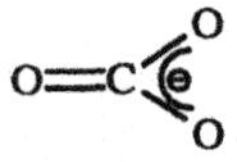   carbonat 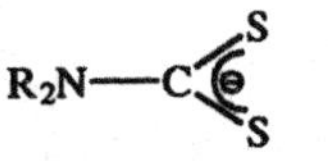   dithiocarbamate 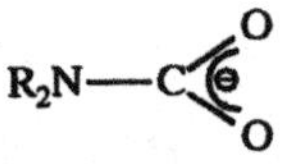   carbamate 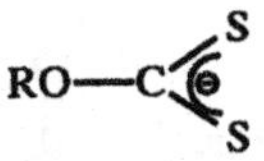   xanthate

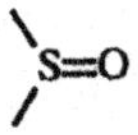

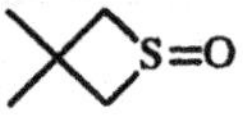

dmto

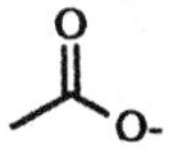

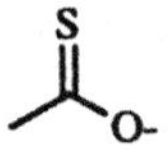

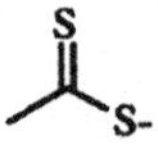

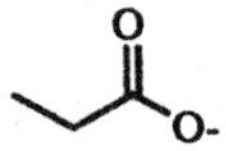

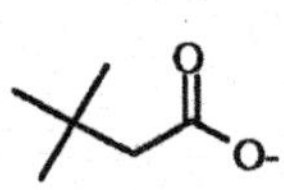

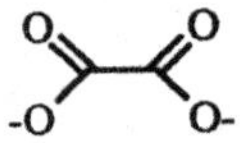

ox 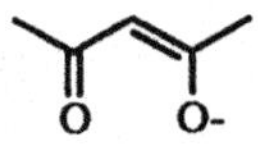   acac 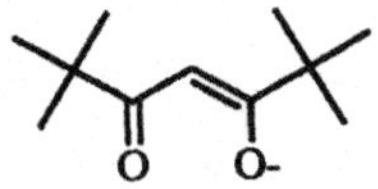   dmp 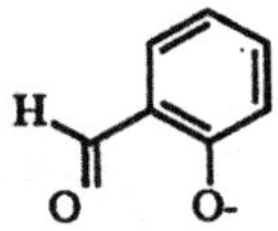   sal

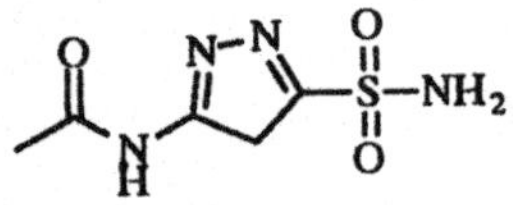

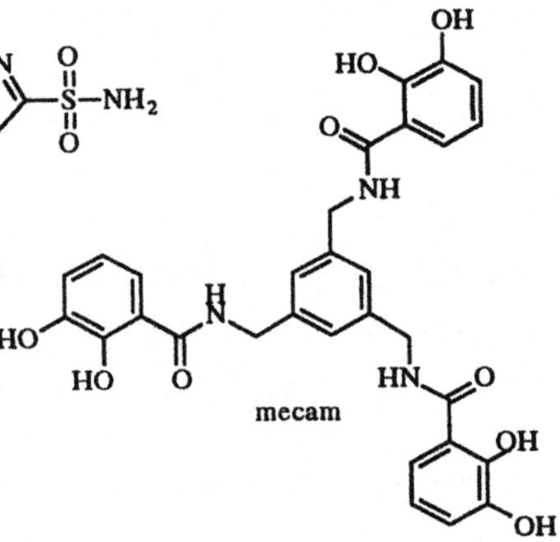

Me₂sal₂tn

mecam

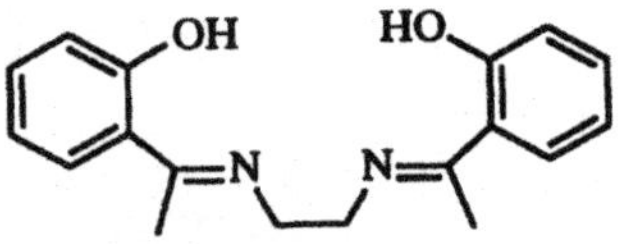

ur 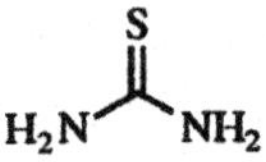   tu 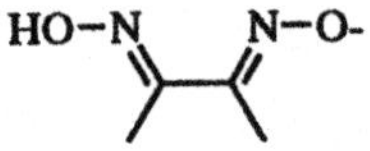   dmg 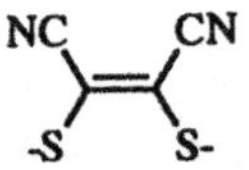   mnt 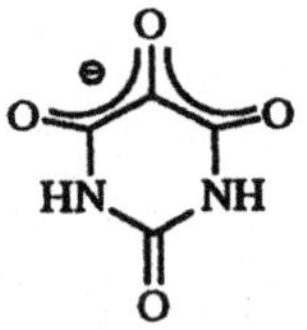   nin 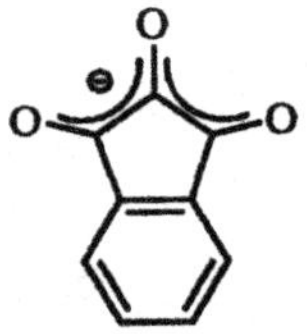   all

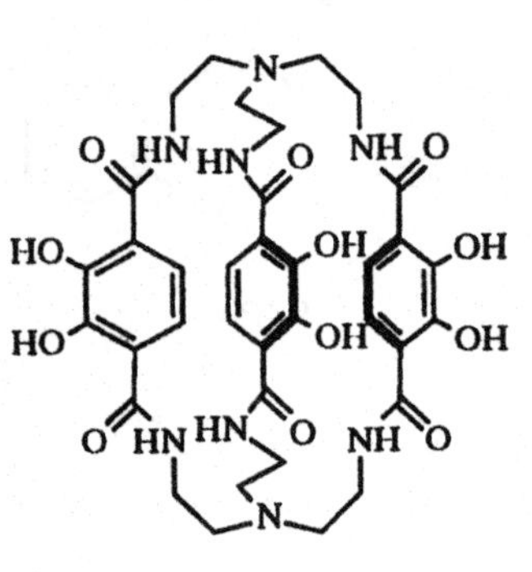

dbp

BCT 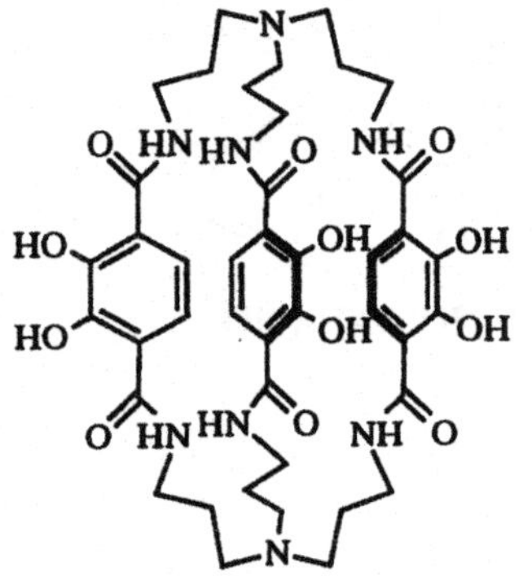   BCTPT

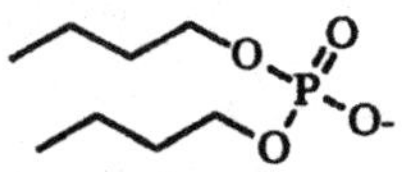

4. P/As-ligands

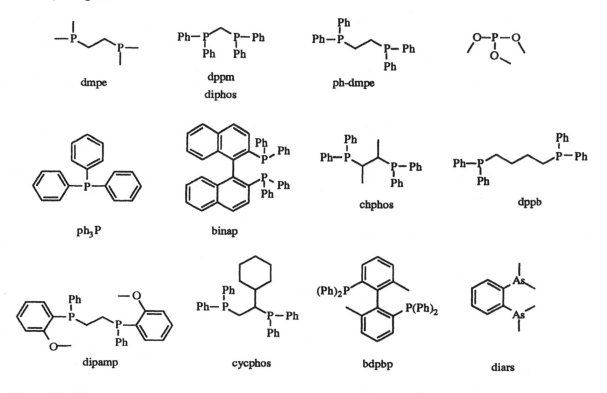

dmpe

dppm
diphos

ph-dmpe

ph₃P

binap

chphos

dppb

dipamp

cycphos

bdpbp

diars

5. π-ligands

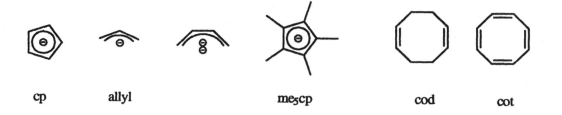

cp   allyl   me₅cp   cod   cot

6. macrocyclic and cage ligands

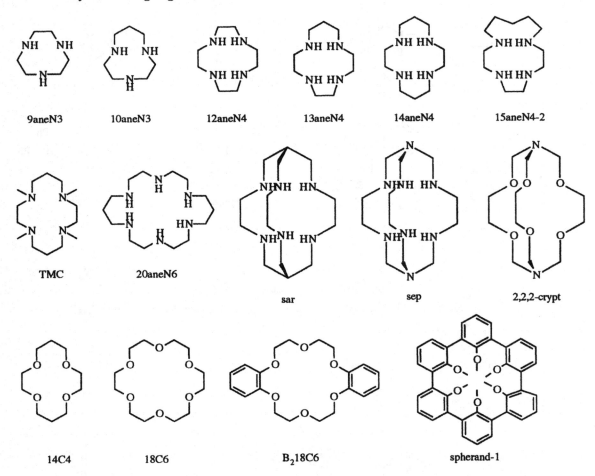

9aneN3   10aneN3   12aneN4   13aneN4   14aneN4   15aneN4-2

TMC   20aneN6   sar   sep   2,2,2-crypt

14C4   18C6   B$_2$18C6   spherand-1

Index